JOURNEYS OF A FIELD BIOLOGIST

ADVENTURES WITH SNAKES & OTHER CRITTERS

Dr. Neil Ford

JOURNEYS OF A FIELD BIOLOGIST

ADVENTURES WITH SNAKES & OTHER CRITTERS

Dr. Neil Ford

UT TYLER PRESS

ISBN: 978-1-7333299-3-4

Published by the UT Tyler Press, an imprint of The University of Texas at Tyler.

UT Tyler Press
Lucas Roebuck, Publisher
Emily Battle, Editor
April Moore, Design and Illustrations
Olivia Paek, Production Supervisor

3900 University Blvd.
Tyler, Texas 75799
uttyler.edu/press

TABLE OF CONTENTS

ACKNOWLEDGMENTS

I would like to thank all the biologists I traveled with, who like me, loved studying animals in the wild: Dr. Harvey Lillywhite, Dr. Charles Carpenter (Doc), Johnny Binder, Dr. Francois Brischoux, Terri Cox, Dr. Jennifer Daltry, Robert Henderson, Dr. Olivier Lourdais, Dr. Richard Montanucci, Dr. Jesús Rivas, Dr. Lynn Sherrod and especially Dr. Rich Seigel. I thank Dr. Marty Crump for sharing her experiences of Costa Rica and helping me develop plans to visit that country.

I thank the various colleagues who allowed me to use their photographs: Harvey Lillywhite, Phillippe Macquet, Bob Henderson, Rich Seigel, Lynn Sherrod, Paul Hampton and Mael Dewynter. Rick Shine, Bob Henderson and Jean-Pierre Vacher reviewed chapters in which they were involved. And thanks to the UT Tyler Press staff for their hard work making this book happen.

My trip to Haiti was supported by The University of Texas at Tyler. The Columbus Zoo and the Cleveland Metroparks Zoo gave funds for my trip to Venezuela. Funding for trips to Trinidad came from the Cameron Park Zoo in Waco. The Rosa May Griffith Foundation funded my trip to French Guiana and supported trips to the World Congress of Herpetology. The Ed and Mary Health Foundation also funded trips to the World Congresses.

I dedicate this book to my parents, Charles and Emily Ford,
who gave me the freedom to develop my passion for snakes.
And to my wife, Joan, who encouraged my numerous travels
to collect animals in faraway places.

I tried to recall if any personal events may have been key in my becoming a herpetologist, and there were several.

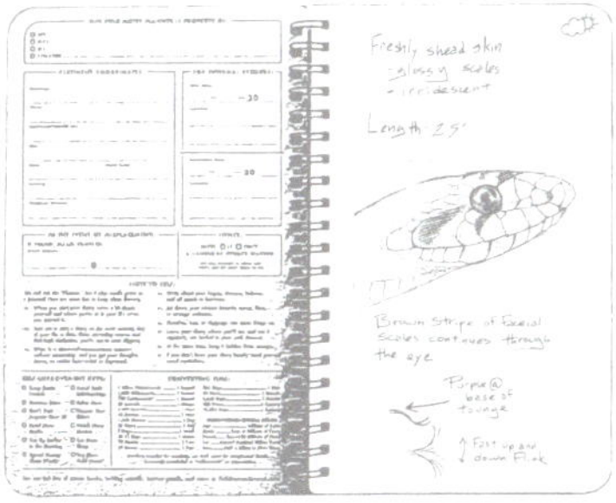

PREFACE *How to Make a Field Biologist*

I have spent over 50 years of my life studying snakes in the field. This book is a compilation of stories about trips I have taken throughout the world, collecting and studying those animals. I am often asked why I became so fascinated with snakes, and I don't have an answer. I have asked my folks if they had any idea, and they only suggested that they "let me be me." During the time in which I grew up, there were no obvious role models to suggest one could make a living studying snakes. I tried to recall if any personal events may have been key in my becoming a herpetologist, and there were several. Interestingly, it appears that almost all of those events were serendipitous. I think of those as "happy accidents" that led me to a career studying the ecology and behavior of snakes. Most of my fortuitous experiences occurred outside, in the natural world, and one goal of this book is to illustrate the way someone becomes a field biologist.

I have learned a number of lessons on my field journeys, and most trips guided the development of my career in some way. The stories I present here describe exciting or funny experiences that occurred, but I also use the stories to illustrate how scientific knowledge progresses in field studies. I believe that some of these ideas may suggest strategies to develop future biologists in the STEM fields, i.e. science, technology, engineering and math. In my 40 years of teaching field courses, I have found that taking students outside is a great way to engage them. As reflected in this book, one of the traits common to both myself and my field biologist colleagues is our enthusiasm for our work.

We all love to tell others what we know about the animals or plants we study. Whether in Haiti, Manitoba or the U. S., I think the curiosity that snakes brought out in the people we met seemed to be universal.

I entitled this book "Journeys of a Field Biologist: Adventures With Snakes & Other Critters" because scientific travel played a major role in the development of my career in biology. I grew up in a small town in the Ozark Mountains of Missouri and spent a lot of time outdoors, in the backyard and surrounding neighborhoods. Television and other indoor activities were certainly available during my childhood but, to me, there were more enticing things to do outside. Watching ants in the ditch in the front yard. Digging up skeletons of animals in the hay of an abandoned barn next door. Walking in the creek down the road. I found those pastimes more interesting than watching Saturday cartoons that were repeated each weekend.

I may have been outside for an hour or so at a time, but my parents seemed unconcerned about my whereabouts. This may be unthinkable today, but in those days, particularly in a small town, it was normal to not know exactly where your kids were. To read a discussion of how that "free range" behavior has changed in today's world, I would suggest reading "Last Child in the Woods" by Richard Louv. For my generation, this behavior may have been a matter of expecting kids to be able to take care of themselves, combined with the fact that neighbors would also be watching. For sure, if any child were doing something wrong, some adult would report it to their mom. The significant point here being that I started taking "journeys" outside early in life.

My dad managed a "five-and-dime" store, and my mother stayed home raising my two siblings and me. My parents certainly had no thought of me becoming a biologist. That career path was unknown to them. Like lots of other folks in Southwestern Missouri, we went fishing and swimming on weekends in local lakes, and it was on one of those trips, when I was about 8 years old, that serendipity struck. I caught my first snake. I have no idea why I was hiking up a dry creek bed and flipping rocks, but my parents didn't stop me. They definitely did not teach me this time-honored snake-hunting technique. That behavioral trait must simply be present in the DNA of herpetologists — although I would guess most children flip rocks to see what is under them, until some adult tells them not to! In my experience, children are full of natural curiosity, especially about snakes. My parents encouraged my curiosity by not suppressing it.

Neil with his older brother, Ron, and younger sister, Cathy.

At some point in my exploration of that creek, I found a worm snake, *Carphophis amoenus*, under a rock. It was a beautiful animal about 8-inches long with a shiny black back and a bright red belly. I think my fascination with snakes was born that very day. I brought the snake back to our trailer, and my dad found a jar for me to put it in for the rest of the weekend. It wasn't until years later that I learned he was very afraid of snakes. I was not allowed to bring the worm snake home, and that was tragic for a boy of 8 or so! However, to answer my many questions, they did buy a book for me, "All About Snakes" by Bessie Hecht. It was published by Random House in 1956, and I still have it on my bookshelf. Later, my high school biology teacher, Mr. Tinsley, who knew nothing about snakes, found me a 1941 "Field Book of Snakes" by Schmidt and Davis. I now knew what a herpetologist was, but not that I could become one. I did convince my folks to allow me to bring snakes home and keep them in a shed in the backyard. The building was unheated, however, and I was 16 before they finally let me keep a king snake in the house over the winter.

Worm snake, *Carphophis amoenus*. (Photo by Paul Hampton)

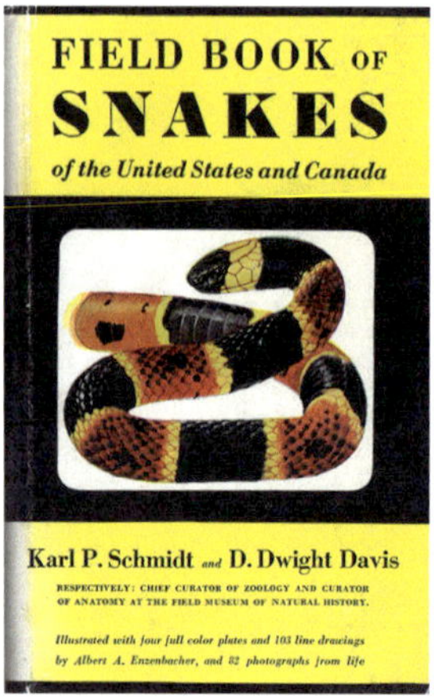

"Field Book of Snakes" by Schmidt and Davis. (1941, G.P. Putnam's Sons)

The Schmidt and Davis book included a figure with the equipment a herpetologist used, but I didn't know old pillowcases could be used as snake bags. I carried my catch home in my hands. After a trip exploring a local cave with my dad, I caught a fat ribbon snake that I thought was pregnant. As I held the snake's neck on the way back to the car, it threw up an adult leopard frog. It was incredible to me that it had consumed such a large meal with such a little head. My dad was just concerned that it was trying to bite me with that open mouth. Later, my mom made bags out of muslin cloth for me to carry my catch.

The first chapter of this book tells the story of serendipitous events — "happy accidents" — that led me to become a herpetologist. The other chapters describe journeys I have taken as a field biologist and my adventures during those trips. Each story also illustrates some aspect of how travel to field sites contributes to the progress of science. Collecting snakes was the goal of each journey, so every story ultimately revolves around how and why I did that.

Snakes manage to do all the survival activities that other animals do — find food, locate a suitable habitat, avoid predation, mate and have offspring — but without limbs. Of the more than 3,000 species worldwide, only a few have been studied to any significant degree and those are primarily snakes of northern temperate regions. This is primarily due to the higher abundance of snakes in those regions (see Chapter 13). I have studied the behavior and ecology of some of the many species living in Northeast Texas, but snakes in other countries were more interesting to me, as they were much less understood. In some of the places I traveled to, I collected a particular species to answer a specific question. Other trips came about by "happy accidents," when I was invited to travel to a particular country by another herpetologist.

A final point I should explain is the kind of research I conduct. Although I call myself a field biologist, my behavioral and physiological studies of snakes were generally conducted in my laboratory in the UT Tyler biology building. Because of difficulty controlling variables such as the weather, conducting research in natural conditions usually takes a long time (although Chapter 13 on the Manitoba garter snakes illustrates an exception). And getting accurate conclusions may take years. For example, after three years of study, I published a paper on the relative abundance of 13 snake species I found in three habitats of Sheff's Wood, a local Nature Conservancy preserve. One conclusion I made in my publication was that the southern copperhead was the most abundant snake in the preserve. However, in the following 11 years, this species was one of the least common.

Snakes are more difficult than most other animals to study in the wild as they forage infrequently, are often nocturnal or cryptic and, in general, are alert to predators. Field studies are critical, though, because the animal's behavior, ecology and physiology in the wild are ultimately what we are trying to understand. Controlled experiments on animals in the laboratory can help increase our understanding of factors that might be difficult to examine in the field. For example, I have recorded mating behaviors of many species in captivity, but I've seen courtship infrequently in the wild, once in anacondas (Chapter 11) and once in garter snakes (Chapter 13). So, although I collected valuable information about snakes in nature during trips, I also brought them back to study in my laboratory. The opportunity to experience exciting destinations while also doing research was just delicious icing on the cake.

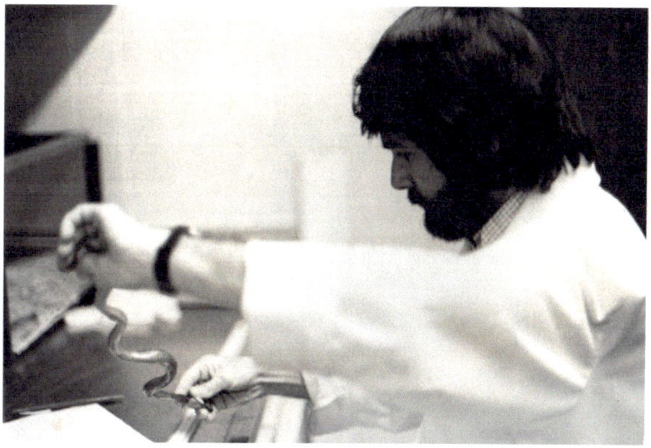

Neil measuring a snake in the laboratory.

When my roommate found out I was interested in reptiles, he introduced me to John — and, in doing so, the incredible idea that one could actually study snakes for a living!

CHAPTER 1

Happy Accidents: My Time at the University of Kansas

I went to the University of Kansas by somewhat of an "accident." In my senior year of high school, my family moved to Kansas City, Kansas. Located only 50 miles away in Lawrence, KU was the logical place for me to attend college. Also, I would get in-state tuition. By serendipity, the university assigned me a freshman roommate whose best friend, John Simmons, worked in the KU Museum of Natural History. John was an undergraduate assistant cataloging preserved frogs for the resident herpetologist, Dr. William Duellman. When my roommate found out I was interested in reptiles, he introduced me to John — and, in doing so, the incredible idea that one could actually study snakes for a living! John introduced me to the scientists in that wonderful museum. I got to hang out in the "Reptile Range" of KU's Dyche Natural History Museum.

The Reptile Range, as it was called, contained the offices of the herpetology professors and their graduate students, with collections of preserved specimens and the workrooms where they examined them. During sack lunches and other gatherings there, the graduate students talked about their research and told stories of their trips to South America and other exotic places. In hearing those stories, I discovered what I wanted to do with my life. "I would become a herpetologist." Many of the students I met in the Reptile Range would become my colleagues for more than 50 years.

Another serendipitous meeting occurred one day as I was exploring exhibits of prehistoric animals found in Kansas that were displayed in the bottom floor of the Dyche museum. As I walked around, I looked through an open door of a side office and saw jars of preserved lizards on shelves. Writing at a desk inside that office was an older, distinguished-looking gentleman. I asked him about the lizards, and instead of saying he was busy as one might have expected, he regaled me with stories of his field studies in China and how the Queen of Siam (now Thailand) did not like his lizards but tolerated him when he worked there. I was enthralled hearing about his travels. Later, I found out from the graduate students how truly important this professor, Edward H. Taylor, was in the world of herpetology. He was the world expert on caecilians, a very unusual group of worm-like amphibians with scales in the skin. Caecilians only occurred in tropical habitats, and Taylor had discovered many previously unknown species on his trips into unexplored jungles. To learn about Taylor's earliest explorations in the Philippines, read his autobiography of that time. It is available at the University of Kansas Museum of Natural History as "Edward H. Taylor: Recollections of a herpetologist," 1975. As a description of how incredibly difficult scientific travel was in the early part of the 20th century, it makes wonderful reading. At the time I met him, I would have been about the age Taylor was when he went to the Philippines. I vowed that day to travel to find snakes in other countries. I so easily could have simply walked on past that open door!

Caecilia tentaculata from French Guiana. (Courtesy of Phillippe Macquet)

My first scientific explorations were not as exciting as Taylor's but I surely enjoyed them. My early journeys were usually just weekend trips with various graduate students from the museum. They occasionally asked undergraduate students like me to help with their projects, and anytime this involved going into the field to collect reptiles, I eagerly volunteered. As luck would have it, most of the undergraduate biology students were premed majors and did not take advantage of these opportunities. So I was invited on a number of collecting trips — much to the detriment of my grades, it turned out.

One trip I remember in particular was with Richard Montanucci (now faculty emeritus at Clemson University) who was studying the banding patterns of northern water snakes *(Nerodia sipedon)*. We collected the snakes by turning over rocks in a river bed, just as I had back in the Missouri Ozarks as a child. But this time, I had someone to show me how to measure and sex them and explain to me why their bands were important. Richard said that when the snakes crawled away rapidly, because of the flicker rate in the visual cortex of the brain of the predator, their bands blended together as solid grey. Then, when the snake stopped moving, the bands appeared again, but now they matched the rocks and the snake "disappeared" in the eyes of the predator (published by Harvey Pough in Copeia, 1976).

The graduate students in the Reptile Range were always willing to talk to me about the work they were doing. Although I was not aware of it at the time, I was participating in one of the most important aspects of being a scientist: making observations and generating hypotheses. They also taught me that reading about my subject in the scientific literature was critical. The graduate students often answered my questions by referring me to a certain published paper to read. I discovered they were also more likely to invite me to go along on trips if I knew a little bit about the subject! However, maybe the most important thing I learned was to take easily-carried, high-energy food (like the sardines Richard brought) on all field trips. Breaks for meals were often short or nonexistent.

I had always explored for reptiles on my own, but now I was learning from the herpetology graduate students in the museum where to look. In the spring, I often visited a den where different snake species came together to hibernate over the winter. It was a forested hillside between Kansas City and Lawrence with plenty of flat limestone rocks to flip. To me, going there was like Christmas morning. Lifting a rock and finding a ringneck snake or a copperhead was like opening a wrapped present. I learned important lessons on that hillside. The leaf-colored copperheads certainly taught me to look carefully *before* I grabbed for whatever was under the rock. A large timber rattlesnake I found there taught me to always bring a really big snake bag on any field trip. Getting a 5-foot-long venomous snake into a 2-foot-long bag was close to impossible and was definitely risky. When I bumped the bag with the rattlesnake, causing the snake to rattle as I entered our house, I also learned that, although I was allowed to keep garter snakes in our basement during the summers, the rattlesnake was *not* welcome in my parents' house.

As spring ended each year, the snakes at the dens dispersed and thus became much harder to find. So, during the summers, I satisfied my need to catch snakes by hunting along a creek near my apartment in Lawrence. It produced an abundance of one species, the red-sided garter snake *(Thamnophis sirtalis)*. These snakes were striped and hard to see in the grassy banks, but when they began moving, the stripes made their movement difficult to follow. It was a different antipredator strategy than the bands on the water snakes. However, I learned to catch the garter snakes by grabbing about 2-feet in front of where they seemed to be. I became intrigued by the diverse colors of the stripes and bodies of garter snakes just within this small area. The variation within these snakes (the genus, *Thamnophis*), both within one species like the red-sided garter and among the 30 other species in North America, became "a major theme" in my graduate student research on garter snakes ("The Garter Snakes" by Rossman, Ford and Seigel, 1996).

Although I helped the graduate students collect animals for their own scientific studies, my field trips to collect snakes were, I thought, just for fun. However, observation is the first step in science and I was doing that as I looked for snakes. Later, as I watched the graduate students take careful

Neil with red-sided garter snake.

and detailed notes whenever we were in the field, I learned how important that was. I received more formal training in that process when I took a course at KU called Museum Techniques. This class was loosely organized around learning various techniques used by biologists in field research. Many of the museum researchers at KU were "Alpha Taxonomists," i.e. they went searching for new species or hunted specifically in areas that had little exploration. This course gave us hands-on experience on techniques used in museums and fieldwork. We learned to trap mammals, how to skin and stuff birds (found dead on the road), how to make resin replicas of specimens and, very importantly, how to keep a field notebook.

Our instructor, Mr. Bee, would take us on field trips and we had to input very specific information in our notebooks. These field notes were written in a traditional way on lined, 100% rag paper — made from fabric, not wood pulp. The notes were kept in a green, three-ring, 5-by-7 loose-leaf binder. Our name and the year had to be placed at the top left of each page, with the date and time in the left margin for each entry. The month had to be spelled out every time because, if it was written as a number, it could be confused with the day of the month. The locality was underlined with a wavy line and had to include the county. The only pen to be used was a rapidograph, filled with permanent India ink. The reason for all of this was so that, if we dropped the field notebook in a lake, we could fish it out, dry out the paper and it would still be readable. I never dropped any of mine in water, but the permanence of those records is evident by the fact that my 1973 notebook is still legible. Indeed, looking over my 40 years of field notebooks to write about my adventures made me realize that many of my memories of the events have been inaccurate.

We often made interesting maps and drawings on the backside of the pages, and with the rapidograph, we could change the pen's tip size from very fine to large lines for shading in dark areas. I drew a map of our small mammal traplines from 1973 (see my drawing of Paradise Point, Perry

Lake, Jeff County, Kansas). With that information and the maps, I could go back to the exact locations where we had placed traps and repeat the same collection again. Indeed, Dr. Fred Gehlbach, a biologist at Baylor University, spent a 12-month sabbatical collecting small mammals along the Rio Grande River, exactly where he had trapped 25 years before. Fred wrote about the changes observed in those two field trips in the 1981 book "Mountain Islands and Desert Seas" (Texas A&M University Press).

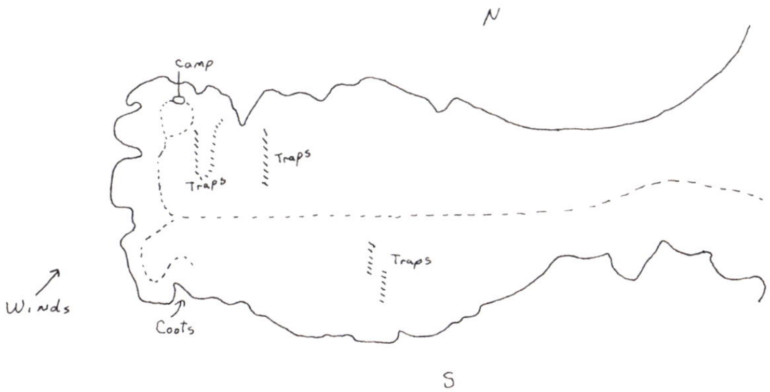

Map of Paradise Point from Neil's 1973 field book.

Another formative class for me at KU was The History of Scientific Voyages. This was the '70s and the university had a "loose" curriculum at the time, so I was able to take this as my required history course. In my biology courses, I had learned about important writings of famous biologists, i.e. the "Origin of Species" by Charles Darwin in 1859. I learned in the class on scientific voyages that early biologists also frequently wrote books for laypersons, describing the travels during which they collected data for their "real" science. We read exciting travelogues including the "Voyage of the Beagle" by Darwin, "Travels on the Amazon and Rio Negro" by Alfred Russel Wallace and the "Naturalist in Nicaragua" by Thomas Belt. I was ready to travel after taking this course.

I left Kansas with $110 in my pocket for the next eight weeks, living on mostly Spaghettios and peanut butter sandwiches.

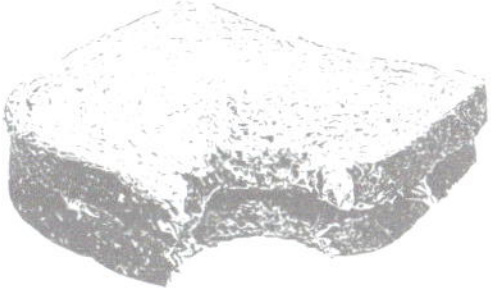

CHAPTER 2

My First Long Trip: The Chaparral in the Laguna Mountains of Southern California

One strategy I adopted at the University of Kansas was to take every opportunity to go on field trips. I kept my ears open to any requests professors or graduate students made for assistants. I had little money to go on these trips but always found a way to make it happen. In 1973, I took comparative physiology from Dr. Harvey Lillywhite and learned he was looking for assistants to spend the summer helping him study the effect of fire on reptiles in the chaparral of the Laguna Mountains of Southern California. So in May 1973, I left Kansas with $110 in my pocket for the next eight weeks, living on mostly Spaghettios and peanut butter sandwiches.

I rode out to Southern California with Fred North, another biology undergraduate, who had helped Harvey the year before. Fred was a premed major with a new Audi, which he refused to drive faster than 55 mph, so we took four days to get to the field site. On the way, we camped in rest areas in New Mexico and Arizona. In Arizona we took the Ajo Way, which was famous among the grad students in the museum as a great road to drive to look for snakes. I had never been through this arid land before, and in my journal, I kept records of all the species of live and DOR (dead on road) snakes we saw.

Fred and I made it to Harvey's study area near Sheephead Mountain in the southern part of the Laguna Mountains late at night, so we camped out at the park and met Harvey at the ranger station the next day. We must have made some prior arrangement to know where to find him, as there weren't cell phones in those days. Planning is still important in fieldwork.

Harvey was investigating how fire changed the composition of snake communities in chaparral. This habitat was adapted to fire, and areas there burned frequently. Because it rarely rained in this chaparral, the dirt roads were dry and dusty, so snakes left trails when they crossed. To see how the types of snake species and their numbers were affected by fire, Harvey compared the number of snake trails on the dirt roads in areas with recent fires to areas that had not burned in 20 years.

Sheephead Mountain in the Laguna Mountains. (Photo by Harvey Lillywhite)

Under Harvey's direction, I developed my own project, which was to look at lizard perch sites in areas that had burned recently. When the burned bark came off the branches of the manzanita bushes, the remaining stems were white, so there were areas here that had both black and white bushes.

The lizards *(Sceloporus occidentalis)* were dark, and we hypothesized that the lizards would prefer the bushes with black branches, as they would be less visible to predators when they perched on those stems. To test this idea, I watched the lizards from the top of boulders for several hours every day. While watching these lizards, I had lots of time to write in my journal and include detailed maps of each study area.

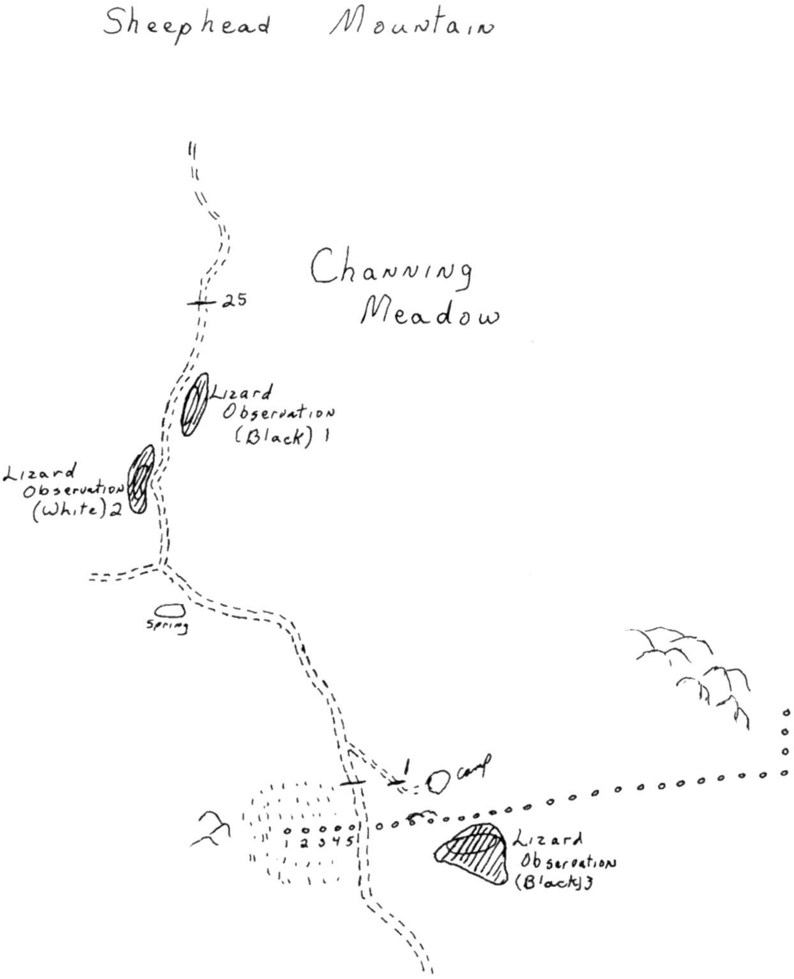

Map of my lizard observation sites.

We also dug lizard pit traps by burying plastic gallon jars that we gathered from local fast-food restaurants and propping flat rocks over them. We placed those traps in areas that were treated with herbicide (2-4-D) and in control areas with no herbicide treatment. The terrestrial lizards like the whiptails, *Aspidoscelis,* would run under the rock to escape predators and fall into the jar. We marked the captured lizards by clipping off claws, and we released them the next day. We compared the number and species we caught in the control areas to those that had been treated with herbicide. Harvey believed the herbicide was reducing the species diversity of lizards.

For the snake survey, we walked three miles of dirt roads in each of three areas: a burned area, an unburned area and a transition area near the edge of the burned site. We walked each area three times a day, every day for 27 miles a day at an elevation of 6,000 feet. I was in great shape by the end of the eight weeks! We recorded any snake tracks that were encountered, and then rubbed them out. Harvey taught us how to tell the species of the snake that laid the trail. Striped racer, *Coluber lateralis,* trails looped widely in a big S-shape as they crawled across the road. Gopher snake, *Pituophis catenifer,* trails were wider and straighter. Southern Pacific rattlesnakes, *Crotalus oreganus,* crawled with the rattle up. Their tracks had a second trail imprint inside the main trail, caused by the elevated tail. (Lillywhite, 1982)

Track of an unknown small colubrid crawling over sand in Borrego Valley, San Diego County, California. (Photo by Harvey Lillywhite)

The striped racers were black with light stripes, so they looked very much like fallen, burned manzanita branches that had bark sloughed off the edges. The Pacific rattlesnakes were also black in color with dark blotches, so they blended in with burned plants. The San Diego gopher snakes were light brown. Because the gopher snakes were more visible in the burned areas, we expected their numbers to be lower in those areas.

Harvey had a VW bus, and after training us for a few days, he began dropping us off to walk different areas of the roads. If we saw snakes while walking and counting trails, we caught and measured them. This included rattlesnakes. We didn't have snake hooks. We used whatever stick was around to pin these venomous snakes. I look back and think it probably was not smart of us to handle rattlesnakes by ourselves with no way to get help. But we were young and invincible. It never entered our minds that this might be dangerous! One of the significant changes in fieldwork today is the required safety training that precedes any research.

It rarely rained at our field sites in the Laguna Mountains, so the roads were dusty and trails were easily seen. After the one sprinkle during our eight weeks there, Harvey bought two push brooms to loosen up the dirt on the roads so the tracks would show up again. That day, two gold miners on motorcycles passed by as I brushed the road with the push broom. I'm sure they wondered what in the world I was doing cleaning a dirt road way out in the boonies!

As I look back through my field notes of that trip, one thing that strikes me is how detailed they were. I carefully collected data on the snake trails we saw each time we walked our part of the road, the measurements of lizards we caught in the pit traps in the herbicide treated area, and which branches the lizards I was watching chose to perch on (this later became my second published scientific paper). My journal also included lots of other things that happened: my observation of a golden eagle soaring overhead as I was setting on a boulder, the time Harvey threw the minibike over a cliff after it broke down for the umpteenth time, and my attempt to pin a huge Pacific rattlesnake with a large branch when I couldn't find a better stick to use. The snake continually pulled its head out from under the branch, so I gave up the effort. Unlike laboratory work, field trips often involve periods of downtime, allowing participants to gather their thoughts, write in journals and just enjoy.

After eight weeks, this adventure was over and Fred and I started the journey home. We drove all day and night in shifts and saw many snakes on the road, including a Hopi rattlesnake, *Crotalus viridus,* just north of Canyon De chelle. In Colorado, we stopped at Long Peak, since Fred really wanted to climb this mountain. We camped overnight at the base of the mountain and got an early start the next morning. It was hot as it was July, so I only took a light jacket. I didn't realize how cold it could be at the top. We started the climb at 9,000 feet. I was in good shape from all of our hiking at 6,000 feet, and the climb started without a glitch. On the "technical" part of the climb, the trail was embedded with cable to help climbers ascend steeper parts. That cable was eventually taken down because too many people fell off the mountain. The metal cable froze my hands as I did not have gloves. We did make it to the top, but then it started snowing. We planned to eat lunch at the top, at 14,000-foot elevation, but we were nearly freezing. Instead, we headed down the "easy" way, slipping and sliding down a big rubble field. We survived this adventure too and resumed our trip back to Kansas.

Relative to my becoming a biologist, two important things came out of my first long field trip. Harvey, Fred and I published the data I had collected on the lizard perch site selection in a scientific paper. My grades at KU were below that required to get into graduate school at most universities. However, since having a published paper was unusual for an undergraduate, I was accepted on probation to the zoology program at the University of Oklahoma (Chapter 3). The second important event occurred when my master's advisor at OU, Dr. Charles Carpenter, asked what I wanted to work on for my thesis. I remembered seeing some strange looking snake trails on the Laguna Mountains trip. These were double tracks. Next to or even within a track, there would be a second trail impression. When I asked Harvey about them, he explained that the second track was from a male snake following, almost exactly, the path of a female. During the breeding season, the female snake produced a chemical called a pheromone and left a trail for the male to follow. I told "Doc" (as Dr. Carpenter was called) that I wanted to work on pheromone trails in snakes and he consented. One of Doc's outstanding characteristics was his willingness to let his students' pursue studies to suit their interests. So I studied the phenomenon of pheromone trails in snakes for the next decade.

“Doc fostered an environment where doing research was fun and exciting.”

CHAPTER 3 *My Time at the University of Oklahoma: A Safari With Doc Carpenter and Field Trips With Graduate Students*

Dr. Charles "Doc" Carpenter was my master's degree advisor at the University of Oklahoma, and although those three years were rough for me academically, as I look back on them I think they had a major impact on my future as a field biologist. First off, Doc did not assign our research projects but rather asked us what we wanted to study. This freedom gave his students the opportunity to develop their own area of research. Sometimes this meant we made mistakes, as I did my first year, but early failures ultimately meant we learned how to do good research. As I look back over the years, I estimate that more than half of my planned research projects failed. Sometimes after years of working on them.

Until years later, I also didn't realize the second important thing I learned during this time: Relationships with colleagues are a key factor for successful research. Many students helped me along the way during my three years at OU, and ever since, I have helped colleagues and students whenever possible. Doc didn't deliberately try to develop altruistic traits in us, but he fostered an environment where doing research was fun and exciting. It was as if we were 8 years old again, flipping rocks in a creek to see what was underneath.

Doc's graduate students in 1980 on a trip to the Southwestern Association of Naturalists meeting in New Orleans. Pictured (from left): Ed and Beth Leuck, Jim Gillingham, Ron Gratz and Neil.

Like other important events in my life, this one also started accidentally. When I asked my undergraduate advisor at Kansas, Dr. Bill Duellman, for recommendations as to where I should attend graduate school, he suggested I contact three people, William Degenhardt at the University of New Mexico, Roy McDiarmid at the University of South Florida and Victor Hutchison at the University of Oklahoma. William Degenhardt's lab was full at the time, so I visited the University of South Florida over Christmas break to check out Roy McDiarmid's lab. He primarily worked on the ecology of local species of reptiles.

Roy picked me up at the Tampa airport and took me to his lab in the zoology building. We didn't stay there long, however, because one of his students was boiling the meat off the bones of a lion that had died at Busch gardens. The skeleton was going into the museum at USF, but the student had gone to a New Year's party and forgotten about the lion. By the time we arrived, the odor from the lion parts left in the pot for three days had permeated the building. So instead of touring the lab, Roy arranged for me to go into the field with Steve Godley, one of his graduate students. Steve took me canoeing on some of the local ponds, and we got the first crayfish snakes, *Regina*, I had seen. Afterward, Roy offered to take me on as his graduate student, but the zoology department at USF did not have funding for graduate students and I had to have a stipend. The University of Oklahoma did offer me funding, so I moved to Norman in August of 1973 to start my master's program in zoology.

I started at OU on probation because my undergraduate grades were low. Taking scientific German and organic chemistry the same semester had not been smart and really hurt my grade point average. The zoology department at OU required a B average to be considered and I had a 2.98. I believe the two papers I published as an undergraduate helped get me accepted, but I also remember that there were fewer candidates applying that year.

When I got to Norman, Victor Hutchison asked if I would like to work with Charles Carpenter instead of him, as Doc had some research funding for his work on lizard behavior and needed a graduate assistant. This offer turned out to be very fortuitous, as Vic was pretty hardcore ex-military who was hired as chairman to "tighten up the ship" in the zoology department. I was close to being a "hippy," so we butted heads several times over the next few years. I don't think we would have worked well together. I became Doc's research assistant, working on a behavior project with him, instead of physiology with Vic. Interestingly, Vic and I have enjoyed each other's company at scientific meetings ever since.

My first year at the OU, I worked for Doc at the university's Biological Station. He taught there in the summer and I started my research project there. He had asked what research I wanted to do for my thesis, and I remembered Lillywhite telling me about snake pheromone trails, so

I suggested that I study those. Doc told me about Harold Dundee's work on snake pheromone trails (Dundee and Miller, 1968) and said to go ahead and design my project.

Doc was very hands-off in terms of his students' work. If we asked for advice or help, he would give it but generally left us alone. Over the long term that lack of oversight was perfect for me, but in the first year I ended up designing some equipment for my research that did not work. So it took me an extra year to finish my project. However, that first failure allowed me to design better experiments, and I became the expert in terms of how to study pheromone trails in snakes.

Because funding for most master's programs today are generally limited to two years, professors tend to give their students very specific projects they know will work. Because the research is so defined, the students rarely have to solve problems and don't always learn how to work independently. This assigning of projects is also driven by the nature of grant funding today. Professors have to produce results within the relatively short time frame of the grant. The zoology faculty at OU were very tolerant of students taking however long they needed to produce good results. I also think this attitude was related to a number of the faculty being field-oriented in their research areas. Field work often did take longer than laboratory projects.

The work I did as Doc's research assistant was to analyze films he took of lizard territorial behavior. He was looking at the evolution of the different displays between related species of fence lizards, *Sceloporus,* that he collected throughout the Southwestern U. S. and Mexico. This research involved measuring head and body movements of the lizard display behaviors frame by frame, with a stop-motion analyzer that Doc purchased on a grant. Each species of fence lizard bobbed up and down in a pattern that was specific only to that species. This display attracted females of their own species and chased away competing males. My job was to convert that movement from the film into a graph for comparing species.

Texas spiny lizard, Sceloporus olivaceus, displaying.

Jim Gillingham, a Ph.D. student, was Doc's assistant the year before I arrived at OU. Following Jim was tough for me because I wasn't nearly as experienced. Jim was studying evolutionary differences in courtship behavior in rat snakes of the genus *Elaphe* (now *Pantherophis*). Bob Ball, Al Coulter and Beth Leuck were other students of Doc's, and Ron Gratz was a herpetology student with Vic. So the six of us spent a lot of time together, hunting snakes in the field and helping each other in the lab. I enjoyed those three years very much because of that camaraderie. I was pretty naïve and might not have made it through without their help and advice!

Field trips at OU were important in fostering my growth as a biologist in a variety of ways. Some field trips Doc planned, but many occurred just because of the type of students he recruited. Although most of their projects involved studying behavior in the laboratory, the animals had to be collected on trips. Therefore, all of his students were "field types." Like me, they loved going out in the field. Anytime someone was doing fieldwork, other students were happy to go along. Many of my journal notes from that time are about local field trips taken with other graduate students.

I roomed with an ornithology Ph.D. student, Joe Grzybowski (now faculty at University of Central Oklahoma), and we took lots of field trips throughout Oklahoma. I was taking an ornithology class at the time, and Joe helped me learn my birds. We often went birding at the university's nature preserve, Oliver's Wood (https://biosurvey.ou.edu/oliver/1966story.html). UT Tyler had a small 10-acre nature preserve on campus much like Oliver's Wood.

I did the most "herping" with Bob Ball, as he enjoyed going with me to collect the ribbon snakes, *Thamnophis Proximus,* that I needed for my thesis project. Some trips were particularly memorable, although I wrote about most of them in my journal. On one night-trip, Bob was bitten by a pygmy rattlesnake he was trying to pin. The snake was not deadly, but the painful bite did cut the trip short. On another night drive, his truck got stuck in a muddy road. We had to walk several miles to a bar and call Jim to pick us up. I paid Jim back by helping him get a monocled cobra out of a wall in the behavior lab! That story I probably shouldn't tell.

Another journal entry describes a walk to the Canadian River south of campus, during which I heard a frog's distress call. I found the frog was being held by one hind leg in the mouth of a blotched water snake, *Nerodia erythrogaster.* Frogs have a distinctive call when they are caught by a snake. The call attracts larger predators — like me — and the snake has to release the frog or be caught itself. I missed the snake that day, but it did release the frog! Since then, I have caught many a snake by tracking down that distinctive distress call!

In my second year at OU, I took field entomology from Dr. Harley Brown, who was an outstanding naturalist. I went in the field often to collect insects for that class. Although Ron Gratz was not Doc's student he took me to several places (i.e. Lake Thunderbird) that he knew of and helped me collect "bugs." Ron was particularly important to me because, during a scientific meeting we attended, he introduced me to Doug Taylor, the professor who would become my Ph.D. advisor. All of the OU grad students were helpful to me in various ways. By example, they taught me that science works best when everyone is willing to share with other scientists the benefits of their insight and experience.

The safari crew: Neil, Jim Gillingham, Al Coulter and Geoff Carpenter.

Carpenter took his students on trips also, and in January of 1974, he took Jim Gillingham and me to the Dallas Zoo. We met Jim Murphy, the curator of reptiles, and filmed various reptile behaviors, i.e. male rattlesnake combat, turtle courtship and combat between two large cobras. I will never forget the two king cobras on the floor of the little back room of the reptile section. I sat on a ladder, watching the keeper trying to get them to fight. It was pretty spooky. I was used to stout pit vipers, but not 10-foot-long skinny elapids! Vipers can strike fast but only with the front part of the body. They are heavy and can be easily manipulated with a snake hook. Elapids, like cobras, are usually more slender and can move up a hook fast. Each keeper needed a second hook to keep the snake under control.

I did not get a hook!

During spring break on March 22, 1974, Doc took several of us on his annual "safari." These trips ostensibly had goals, and this time, the purpose was to get lizards for Doc's research. But to a large degree, Doc took trips because he enjoyed the time with the students. Maybe he also realized the time we spent together was important in helping us form long-term relationships with each other. Doc planned a trip to West Texas this time because Mexico was shut down. I rode in Doc's Volvo station wagon along with his son Geoffrey and Ph.D. students Al Coulter and Jim Gillingham. We camped the first night at Monahans Sandhill State Park in Texas. We then headed for the Davis Mountains and set up camp at Fort Davis State Park. I caught my first checkered garter snake, *Thamnophis marcianus,* at Limpia Canyon in the Davis Mountains — not knowing at the time how important this species would be to me in years to come. One evening, Doc fed us "sow belly and beans," one of his field trip traditions. I didn't deliberately imitate Doc's "safari" when I started teaching. But looking back, I must have enjoyed the experience

Checkered garter snake.

such that I also had spring break field trips most years. I also had a tradition of feeding the students something they generally hadn't tried. In my case, it was sardines on crackers for lunch.

I realized over the years that camaraderie among the students was built most strongly during difficult times — canoes tipping over, violent rainstorms and even accidents. On Doc's "safari" we had a wreck in Del Rio as Doc didn't see a car in a crossing street. We had to stay in a motel playing cards for two days while the car was repaired. It was also hot in West Texas, and we weren't finding many reptiles. So when we were back on the road, we headed through South Texas and then over to Louisiana. We set up camp in Southern Louisiana at Longfellow Evangeline State Park, and by flipping trash around the local bayous, we did catch lots of snakes. My favorites were the speckled king snakes, *Lampropeltis gettula.* Their spots were orange, unlike the white-spotted Oklahoma king snakes. I also collected my first green water snakes *(N. cyclopion),* buttermilk racers *(Coluber constrictor)* and *Amphiuma* (huge aquatic salamanders). I was very impressed with the abundance of snakes in the swamps of this state! On the way back to Oklahoma, we stayed at Louisiana's Kisatchie National Forest, where I did fieldwork 40 years later.

Field trips at OU often occurred in conjunction with driving to scientific meetings. For example, the field entomology class I took went to the Oklahoma Biological Station to attend the Southwestern Association of Parasitologists meeting and also to collect insects. We would camp at parks and recreation areas near a meeting site, and this time we stayed at Bucombe Creek Recreation Area. After the meeting, the entomology class stopped at the museum at Platt National Park. Another time, several of us headed to the annual Southwestern Association of Naturalists meeting at Baylor University in Waco, Texas. On that trip, Jim Gillingham and I camped at a city park near the university, and after the meeting, we went on the sponsored field trip to Devil's Elbow.

Field trips to meetings did build camaraderie among the students, but making contact with professors was also important. During my last year at OU, Ron Gratz and I attended a herpetology meeting at Tulane University in New Orleans. One night there at a bar on Bourbon Street, Ron introduced me to Douglas Taylor, a professor at Miami University in Ohio. While drinking a "hurricane," I told Doug about my research and he suggested I apply at Miami to work with him, as they still had two Ph.D. fellowships available. Ron attended school with Doug when he was doing postdoctoral research and liked him. Ron felt I would be better off leaving OU for my Ph.D., so I applied at Miami. I didn't get a fellowship, but one regular teaching assistantship was available. I was given that my first year, and later I did get both a teaching and research fellowship.

So in 1976, I packed up my little Volkswagen hatchback and drove to Oxford, Ohio, to start my Ph.D. During the next three years I took many field trips collecting garter snakes for my research. And although I detailed lots of those trips in my notebooks, I am going to skip those years, since the only really exciting journey was one I took with Randy Layne to Louisiana to buy live crayfish for a neurophysiology project. Most other trips involved flipping trash around abandoned houses. Unless you are really into catching lots of garter snakes, those trips were not that exciting.

The excursion was successful ... But it was a rough journey.

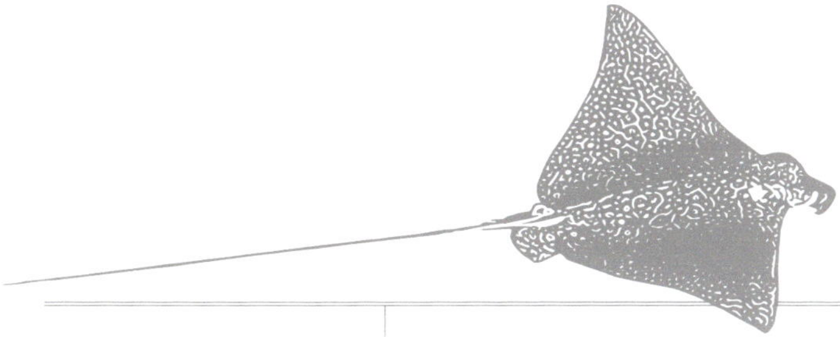

CHAPTER 4

Dictator 'Baby Doc' and the Tonton Macoutes: A Trip to Haiti

In 1982, I went to Haiti to collect three species of vine snakes. I was going to study how the foraging behavior differed among the species. Comparing behaviors in related species is a common way to try to understand the evolution of behavior. However, studying behavior in the wild is difficult. This story in particular illustrates how testing hypotheses generated by field observations can be more easily accomplished in the laboratory. It also presents some of the difficulties of working in foreign countries.

Bob Henderson of the Milwaukee Public Museum worked on islands in the Caribbean. He was specifically collecting in locations where our knowledge of the reptile species living there was limited. During his work on Hispanola (Haiti and the Dominican Republic), Bob noted that, even though there were three species of vine snake (genus *Uromacer*) on the island, he would only find two of the species in any one place. One species (*U. catesbyi*) had a blunt nose and was always present at each site. However, only one of the other two species, both of which had pointed snouts, would be present. The brown species *(U. frenatus)* would be in one area and the green species *(U. oxyrhynchus)* in another. He never found these two species together and hypothesized that they were in competition for the same prey. Bob then documented the diet of the three species in the field by palping out food from the stomach of those he caught. He also looked at the stomach contents of preserved snakes in various museums around the U.S.

In 1987, Bob wrote about that work in a paper entitled "Food Habits of Three Colubrid Tree Snakes (Genus *Uromacer*) on Hispaniola." He found that the two pointed-nose species tended to eat mostly lizards, whereas the one blunt-nosed species also ate tree frogs. He suggested in the paper that the two narrow-snouted species might have better depth perception since the visual fields of their eyes overlapped more. Bob's field observations suggested these two species slowly stalked their prey until they were close enough to strike. So "knowing" the distance to the prey would be important for a successful strike in these snakes. The blunt-snouted species *(U. catesbyi)* would be less able to determine distance to the prey, and Bob suggested they foraged by chasing the lizard or frog. However, Bob rarely saw vine snakes foraging for prey in the wild. I suggested to him that it would be easy to document this behavior in the lab to test his idea. So we decided to go on a collecting trip to Haiti, and Bob arranged permits for me to bring live snakes back to the U.S.

Therefore, in late December of 1982, I headed to Port-au-Prince, the capital of Haiti. This would be my first field trip outside the United States. The excursion was successful in that I was able to bring back vine snakes of all three species to study in my lab. But it was a rough journey. I was unprepared for the experience of working in a country like Haiti.

Haiti was, and still is, considered the poorest country in the Western Hemisphere. That crushing poverty dramatically affected most aspects of this trip. I was also naïve about working in a country with an absolute dictator like Jean-Claude "Baby Doc" Duvalier and his secret death squad, the Tonton Macoutes. This name refers to "Uncle Gunnysack," a mythological bogeyman who kidnapped children and put them in a sack to eat later. The Macoutes were given the name after people opposing them began to disappear. This secret police was started by Jean-Claude's father,

"Papa Doc." The Macoutes stoned and burned his opponents alive, hung up corpses and even extorted plasma from the poor to sell to the U. S. They could have killed us too and dumped us in a ditch with no consequences, but I was blithely unaware of this.

The most successful field trips required knowledge of the area to be explored. That required either going with someone familiar with the land or taking two trips — one for scouting and making contacts with locals and the second for the actual trip. Bob was familiar with the unique problems of working in Haiti and gave me a lot of advice before the trip. However, nothing could really prepare me for the magnitude of the problems of overpopulation and poverty in this country. So, although this story presents my adventures collecting animals in Haiti, it also includes how working in this country affected me at the time and changed my understanding of the world beyond the U.S.

My first mistake on this trip happened the day I arrived at Port-au-Prince in Haiti. My luggage didn't make it. I had put my gear in a heavy wooden box and padlocked it. Airlines even in those days wanted to know what their passengers were carrying, so I'm sure they had to check out my locked box. On future trips, I always made sure my luggage could be easily opened, even if it made my equipment more susceptible to "walking off."

Bob and undergraduate students Dan Nedrelo and Jim Krantz from the Carroll College (now Carroll University) met me at the airport. Bob had rented a car, and after finding out my luggage didn't make it, we drove into the city center and stayed overnight at the La Griffone Hotel. As this story goes on, you will see the second lesson I learned. Rental cars in other countries have lots of miles on them and frequently break down. You must be prepared to fix them.

John Thorbjarnarson from the Florida State Museum came over to La Griffone that first evening. He was also working in Haiti and helped with our arrangements. Although I lost track of the two Milwaukee students over the years, I ran into John years later while hunting anacondas in Venezuela. Sadly, he died of malaria in 2010. In the countryside of Haiti, we often saw warning signs of the presence of malaria. I wondered if John contracted the disease there. I took a malaria preventative for the first part of this trip, but stopped taking it because it made me very nauseous. Malaria medicines were hard on the body, so most herpetologists working in areas with mosquitoes could not take them long-term. I know several who suffered the effects of the disease later in life. I also took a prophylactic stomach medicine my doctor recommended, as *E. coli* was known to be a common problem in Haiti. I guess it helped. We did watch what we ate but, of the four of us, I was the only one who did not have stomach issues during the three-week trip. Thanks, Doc.

The next morning, I drove the car back to the airport to see if my box and duffle bag were in. Driving the streets in Port-au-Prince was a crazy undertaking, with motorists rapidly dodging in and out of traffic and no one paying any attention to street signs. As it turned out, the same crazy driving was common in most other less-developed countries I visited. You had to be fearless to drive in any big city. One vehicle unique to Haiti was the Tap Tap. They were small, brightly painted buses or vans with open sides allowing people to give their money to the driver as they jumped on. When someone wanted off, they got the driver to slow down by tapping the top of the bus. Hence the name, Tap Tap.

A Tap Tap. (Photo by Richard Sajdak)

My luggage still hadn't arrived at the

airport, and Bob wanted to get started collecting. We headed to the downtown Iron Market so I could buy enough clothes to last a couple of days. Most bigger cities in other countries I've visited had open-air markets like this, but it was my first experience with one. It was pretty hectic, but the "name brand" clothes sold there were very cheap. Since the car was small for the four of us, we dropped our extra gear — mostly jars, formalin and alcohol — at the U.S. Agency for International Development office in Port-au-Prince. We then headed out of town toward the southwestern peninsula, Tiburon.

Bob planned to collect animals on various islands off the Tiburon peninsula that had not been surveyed for reptiles. Traffic on the main road out of the city was heavy but thinned out as we drove further into the countryside. It was a long drive and we had a late start, so we stopped and camped one night on the beach. We put up one tent for Dan and me. Bob and Jim just slept in the car. Luckily, we were behind some trees and not visible from the road. Otherwise, we probably would have been visited by the Tonton Macoutes.

One island that we surveyed was Ile-a-Vache, the largest island off the peninsula on the southern side of Haiti. We drove to the port town of Les Cayes and inquired at the wharf about boats to take us across the bay to the island. With limited funds, we hired the cheapest sailboat for $20. We should have paid more! We waded out to the boat with our gear on top of our heads. The wind was against us so Justin, the captain, had to tack back and forth to get us across. We had to hit the deck every time the boom swung around, which he found quite amusing. Justin also brought a young boy along to bail the whole way. The waves were high and water poured into the boat.

When we made it over to Ile-a-Vache, we found mangroves surrounded the island. Mangroves are shrubs in the tidal zone along tropical shores and have tangled roots supporting the trees. The captain trimmed the sail and poled the boat along a narrow passage through the mangroves. The mast kept getting hung up in the trees, but finally, the canal widened to a pool that had two other anchored boats. One fisherman had just speared a spotted eagle ray and gave it to Justin to hold up so I could snap a picture. I painted that scene some 30 years later. Again, we had to carry our gear on our heads while wading into the shore. From there, we walked up a narrow path to a clearing with the houses of two families. We arranged with them for Dan and Jim to set up camp. We explained as best we could that we were wanting to get reptiles and would pay for them. We left Dan and Jim with the gear they needed to preserve specimens. We had some misgivings about leaving them there alone, but I needed to get my luggage and Bob wanted to start collecting. We arranged a time to meet up at Les Cayes. Bob and I then

Justin with spotted eagle ray.

Bob unloading on Ile-a-Vache.

took the leaky sailboat back and started the long drive to Port-au-Prince.

The next morning, the porter at Hotel La Griffone told me Eastern Airlines had called. We felt like there was a good chance my luggage was in. After another crazy drive to the airport, I picked up my luggage and went through customs. No planes had arrived recently, so security must not have had much to do. They pulled everything out of my bags. But I didn't mind. I was just happy to get it all. Bob and I drove back to the USAID office where our extra belongings were stored. We dropped off some of my extra gear. Jim Keach, the director, was there this time and told us President Duvalier — Baby Doc — had declared a holiday and was driving around town throwing money out of his car. However, a car bomb had gone off near the president's palace, killing two Haitians and one American. This was my first real inkling that things were a little dangerous here!

Bob and I headed back on the southwestern highway to Les Caye to pick up the guys. It took a lot longer this trip, as we were stopped five times by the Tonton Macoutes, who thoroughly searched our vehicle. We showed them our collecting permit, but they wouldn't even look at it. Haiti had had a written language, Creole, for only 20 years or so. Perhaps they could not read the permit. I had brought along a guidebook to Creole that had just been written, but it was not much help. Every time the Macoutes stopped us, we had to take out all our gear and they always checked carefully under the hood of our vehicle. They apparently were looking for a car bomb.

We arrived at the wharf where we had rented the boat and were immediately surrounded by people. A policeman showed up on a bicycle and we assumed he was a Tonton Macoute, since he was in the tan uniform we saw on those police who had stopped us. We tried to explain why we were there but communication was challenging. Lots of Haitians speak French, but Bob and I didn't know that language either. We finally realized the officer knew what we were doing there, because he asked us about Jim and Dan — and used their full names. He must have gone to the island and checked out the boys' passports.

We finally understood that the boys would be back by noon, so we bought Cokes at a nearby hotel while we waited. When we returned, the guys had arrived and were somewhat worried when we weren't there to meet them. We gave them the drinks we bought for them, and they forgave us. I tried to get a photograph of the Tonton Macoute, but was not going to ask him to pose. So all my photos are from behind, as he rode away on his bicycle.

Jim and Dan told us the village was actually a little bigger than the two huts we had seen. They also said they were surrounded by people at all times, even when they were sleeping. Jim liked to eat and told us the locals ate better than the boys did. While Jim and Dan made do with peanut butter sandwiches, the locals dined on lobster and fish. The second day, the guys were inundated with animals gathered up by the local children. However, most of the lizards were killed with rocks and were not in great shape for

Neil with vine snakes.

The Camp Perrin campsite.

preserving. Surprisingly all the vine snakes were brought in alive. We had heard that Haitians believed snakes could take their souls if touched.

We drove a well-packed gravel road to the tip of the Tiburon Peninsula and spent several days collecting in the Camp Perrin community. We pitched our tents in the parking lot of a small store. It doubled as a house, like most stores in the area. Visitors knew it was a shop only because of the Pepsi sign out front. Madame Jean, the owner's wife, could speak a little English and spread the word that we wanted reptiles and were offering one dollar for each snake. We tried to go out looking for ourselves but found it difficult. Groups of children, ages 10 to 15 or so, followed us everywhere. Once they saw us flipping rocks, they would swarm over the rock piles and tear them apart. The farmers had piled up mounds of rocks as they cleared the fields for planting, accidentally creating great lizard habitat. The kids called the lizards "mabuya" and laughed when we grabbed the ones they found among the rocks. But after we paid a dollar to a young boy who came in carrying a dead vine snake *(Uromacer frenatus)* with a bit of twine around its neck, things changed. They began to search the scrubby acacia trees of the surrounding forest for snakes.

Soon, another boy brought in a ground boa *(Tropidophis)* he found in a rock pile, and then boys came in with more and more snakes. Madame Jean brought out chairs for the front porch and we sat and drank Cokes while the kids brought us animals. It was very pleasant and the easiest "collecting" I ever did. This went on until late morning, when we saw some folks walking down the road carrying the body of a 10-year-old girl, wrapped in a sheet. Lots of the kids went off with the adults then. We didn't find out how the girl died, but we occasionally saw people with polio. Malnutrition also was evident. Considering the condition of the drainage ditches, cholera was probably a recurring problem as well. In all my other trips, I have never seen poverty as severe as in Haiti.

As it approached dusk, Dan and I managed to go hunting without an entourage. We got a Haitian black racer *Hypsirhynchus parvifrons,* which are striped like garter snakes, and also caught several more *Uromacer,* along with a Haitian boa, *Chilabothrus striatus.* The boa had the smelliest musk I had ever encountered in a snake! When we returned to the campsite, a party had started in a nearby shelter. It turned out to be another holiday, and a marimba band started playing. They hooked up a generator for the amps, so it was loud. Some of the more curious party-goers came over to check out our snakes. Showing off snakes has always been a great way to make new contacts.

That night, Dan and I slept in one of the tents but made the mistake of taking off our shoes inside the tent before hitting the sack! I remember that same problem when I worked in the Chaparral of Southern California with Harvey Lillywhite. We aired out the tent in the morning, while trying not to let mosquitoes in. We had seen signs warning that malaria occurred in this area.

As in most tropical nations, the people in Haiti woke early. And apparently, the store where we camped also served as a bus stop. There was a constant flow of people going down the road with goods and packages. So for breakfast, I bought a dozen bananas that one woman was carrying on her head and a loaf of bread from another woman walking by. Madame Jean kindly brought out a cup of coffee for me, as the other guys didn't drink

Ladies carrying bread at Camp Perrin.

coffee. The cup was half full of a thick black liquid. She held up a can of condensed milk, asking if I wanted some in my coffee. It was great! About the equivalent of four cups of coffee in one. Ever since that trip, I have preferred my coffee strong with lots of cream.

I bought a can of pear nectar, and Dan and I went off bird watching. We saw a thick-billed tody, several species of hummingbirds, parula warblers and other birds we couldn't identify. I didn't have a very long lens for my 35mm camera and couldn't get any good photos of birds.

When we returned to the campsite, the kids had brought in lots of snakes and lizards, and Bob and Ned were frantically preserving them. We pitched in to help. Due to the heat, that afternoon was quieter, but it picked up later in the day as the animals started moving again. The kids brought in lots of dwarf boas *(Tropidophis)* and another racer *(Hypsirhynchus parvifrons)*. Then, two older boys brought in a 5-foot boa *(Chilabothrus striatus)* and wanted two dollars for it. It had a large lump midbody, which we explained was likely a rat. That night, we took it out to the forest and palpated out the rat to record the boa's diet. We then let the snake go, as it was too large to preserve. Afterward, we hunted in the scrub for several hours but only found one vine boa *(Chilabothrus gracilis)*. We lost our way at one point but eventually found the road and hiked back to camp.

The next morning, the bus loaded up again at 4 a.m. with lots of

Vine snake, *U. oxyrhynchus*.

Haitian Boa, *C. striatus*.

honking and shouting. We managed to get back to sleep after it left. When we finally crawled out of bed, we discovered our car had a flat tire. A boy was standing nearby, holding an air pump, and was willing to air up the tire for money. Our tires went flat a lot, but I guess we helped enterprising youth. By the afternoon, the kids had brought in over 50 snakes and we told them we didn't need anymore. However, they kept bringing them, so we lowered the price to one gourd (about 25 cents) and most decided that wasn't worth the effort. An English-speaking man rode up on a motorcycle, and we asked him to thank the store owners for letting us stay there. When we told him we visited Ile-a-Vache, he said a missionary lived there with a motorboat and would have given us a quicker and safer ride over to that island! Next time, we will pass on the sailboat!

We headed to the north side of Haiti next. The terrain was much more desert-like with large cacti, and it seemed like fewer people walked the road in the heat. Stopping for a Pepsi at a little store in the town of Camp Coq, we put the word out that we wanted snakes and would be back the next day. Then we drove to Cap Haitian. There was a checkpoint gate into this city, and the police examined our passports as we entered. We saw a young couple hiking on the road and asked if they knew a cheap place to stay. They were staying at the Hotel Universal for $12 a night, so we packed them in the car and drove there to check it out. The beds looked OK and the hotel had showers in a communal bathroom, so we booked rooms. The couple recommended the Hotel Rus Christopher for supper. There was a live band there, and I remember the group singing "Guantanamera" — changing the words to "one dolla billa." It was a sad commentary on the economic situation in Haiti.

The next day, we drove back to Camp Coq and stopped at the same little store. Bob remained there while the rest of us crossed the road and hunted in the banana and orange groves. About 15 urchins followed us around, but all they found was a Haitian boa in a hollow tree. We didn't really want it but gave him a dollar anyway, because the kids would kill it. We released the snake later, down the road. Back at the store, Bob received two *Hypsirhynchus,* but had hoped to get a *Celestus* (a large skink called a galliwasp). So we decided to head over to Dondon — another location where *Celestus* had been recorded to occur. We parked at a bridge, over a dry riverbed, and showed folks nearby our photograph of the lizard. Bob waited in the car while Jim, Dan and I walked down to the riverbed to look for snakes. It was soon evident to us that the dried-up river was the local toilet! Yuck. The kids laughed at us walking through the area. We caught one *Hypsirhynchus* and a marine toad but couldn't take walking in the filthy creek for long. We left the toad in the riverbed. Sadly, we saw the kids stoning it as we walked back to the car. For lunch, we bought oranges and mangoes that were delicious, but we were discouraged about our lack of success. We went back to the Hotel Rus Christopher for dinner so Bob could call Al Schwartz, a herpetologist at Miami-Dade Community College. Al had collected in Haiti, and Bob wanted to see if he knew of more sites where *Celestus* could be found.

Al told Bob we weren't offering enough money. So the next day, we journeyed to a spot north of Camp Coq and asked lots of folks about getting the lizard for us, offering three dollars each. We decided to give them a day to collect, and spent that day at the hotel. That evening, the music continued late into the night. Maybe it was Saturday, I had lost track of the days. When we returned to Camp Coq the next morning, it was like no one remembered we had asked them to collect lizards! More discouraged, we drove up to Cormier Plage, a nice beach on the northern coast of Haiti. I got to use my face mask and snorkel and saw some beautiful marine fish. Another interesting sight: Many of the women on the beach were sunbathing topless.

A dry creek bed, aka toilet.

Neil and Jim on Cormier Plage.

When we were ready to head back to the hotel, we found we had yet another flat tire. Naturally, a boy was nearby with a pump to fix it. It was pretty obvious what was happening. Jim volunteered to sleep in the car that night to keep it safe, but we convinced him it was not worth it. Most of the people we met were obviously extremely poor. Even though we had limited funds, paying someone a dollar to pump up a tire was not a big problem. We saw beggars who were crippled with polio, even though the vaccine had been out since I was a child.

On our way back to Port-au-Prince, we drove past Petion-Ville, an affluent suburb with nice homes, in the hills above the city. The mountainsides were stripped of trees and terribly eroded. When it rained,

mudslides occurred everywhere. In lower parts of the city, there were rows and rows of cardboard shacks that must have been under water during the rainy season.

We stopped by some art galleries on the way back. Haiti was well-known for its somewhat primitive but very colorful style of paintings. However, the paintings were too expensive for us! We also visited a biologist Bob knew at one of the USAID offices. Jim Talbut told us about the current political situation. He said few people spoke out against the dictator because of the myth that Baby Doc could use Voodoo on them — and also because the Tonton Macoutes would kill them. However, it turned out that only three years later, in 1986, a popular uprising overthrew Baby Doc.

My last adventure on this trip occurred at the airport. As several ladies checked my baggage by hand, Tonton Macoutes in their brown fatigues looked on, holding their machine guns at the ready. I had permits for the snakes that were packed in my carry-on backpack, but I am pretty sure they wouldn't have cared. Luckily, I packed really dirty clothes in with the bags

Haitian paintings on wall in Petion-Ville.

containing the snakes, and I guess the ladies didn't check carefully. Whew. I learned to always put dirty laundry in the luggage with the snakes.

Back at my laboratory at UT Tyler I had a graduate student videotape the foraging behavior of the three species of vine snakes. Both of the pointy-snouted snakes stalked prey in the same way. They would weave back and forth like branches in the wind until they came within inches of the lizards, before they would strike. The blunt-nosed species started chasing their prey from several feet away. Bob's hypothesis was correct!

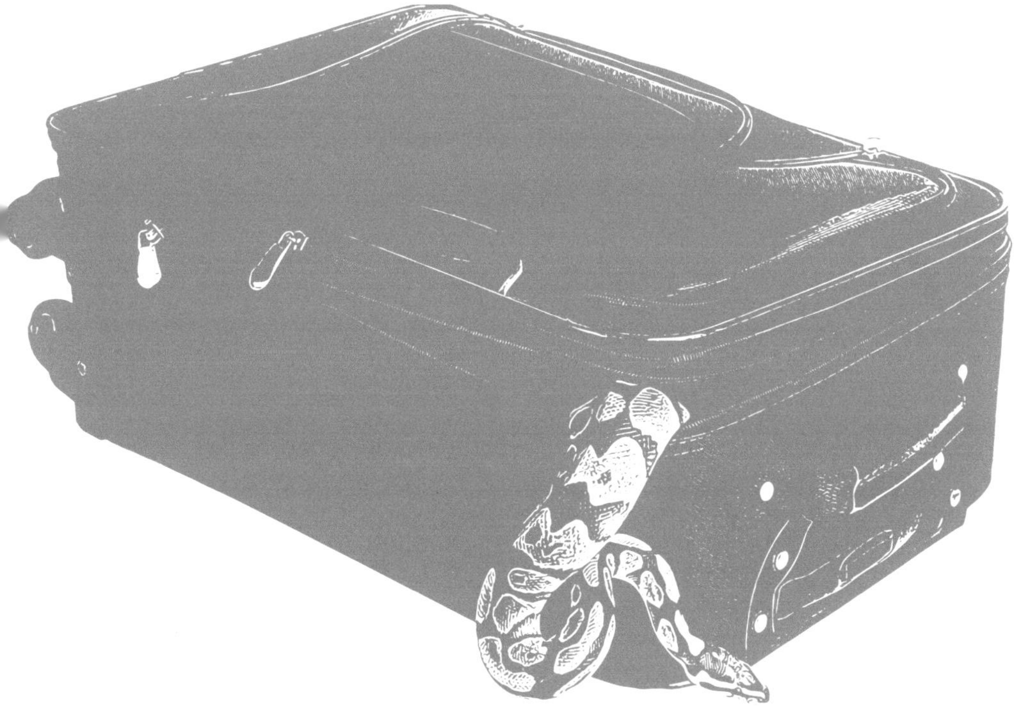

"I took my early graduate students there and told them it was like Christmas morning as you didn't know what you would find."

CHAPTER 5

Trips to South Texas: Little Dramas and Learning From My Students and Volunteers

My Ph.D. involved studying the ability of male garter snakes to follow chemical trails left behind by females of that species. I worked with garter snakes because there were over 30 species and most were common, sometimes in huge numbers when they came out of hibernation dens in the spring (see Chapter 13). I wanted to continue my studies on snake pheromone trailing when I moved to Tyler, Texas. However, there were no species of garter snakes occurring in the immediate area. I called Craig McIntyre, an amateur herpetologist who lived in Austin and knew of populations of garter snakes there. He told me checkered garter snakes, *Thamnophis marcianus,* were common just south of him in Cotulla. Actually, Cotulla was about two hours south of Austin, but I learned that distance in Texas was relative. Even though it was an eight-hour drive from Tyler at the 55-mph speed limit of the time, I took lots of field

Checkered garter snake.

trips to Cotulla over the next few years. This southern species of garter snake did not aggregate in dens like those up north but was still abundant on roads at night in the lower valley of Texas.

Cotulla ended up being my favorite place to hunt snakes in Texas. I went there for the checkered garter snakes, but the area also had lots of other interesting animals. Driving around Cotulla at night, one could reliably encounter large numbers of a variety of snake species. I took my early graduate students there and told them it was like Christmas morning as you didn't know what you would find. Craig came along many times and introduced me to Alan Tennant, who later wrote a book on the "Snakes of Texas." I took my vertebrate natural history classes down there on campouts, not just for the snakes but also to see the many Mexican species of birds there. Over the years, I also took several volunteers from my lab there. Although I brought all these folks along to help me find animals and for the students to learn biology, I also learned a lot from them on those trips. This chapter explores the value of interactions with different kinds of people.

My graduate students in my earlier years came to study snake behavior and ecology. Our master's program, like most others, only funded the graduate students for two years. Conducting work with snakes in the field in that short amount of time was nearly impossible. So my strategy was to find a species common enough so students could collect a sufficient number of individuals to bring back to the university lab. The eight-hour drive to Cotulla was worth it, as we collected a dozen or so adult checkered garter snakes each time. My first two students used these animals to study reproductive behavior in snakes.

Carolyn Schofield (Bronston) was my first grad student. Carolyn had been a high school biology teacher in Houston before moving to Tyler. She came by the university to see about doing a master's degree in biology, and I happened to be the only faculty in that day. I spent some time telling her about my research, but more than anything, I think my enthusiasm for my

Carolyn Schofield (Bronston) at the Cotulla KOA.

work convinced her to request me as her advisor. Carolyn had an intense interest in learning anything in biology. Over the next two years, we took several trips to collect the garter snakes for her project. On those long trips, we had many a great discussion about biology and how to teach it. Her strategies for teaching high school students were quite useful to me in teaching nursing classes that were my "bread and butter" at the time.

On one trip I wrote about in my journal, Carolyn and I camped at Honey Creek Preserve near Austin before we headed to Cotulla. She had brought a Greenburg smoked turkey! I learned from her that we didn't have to eat cheap food on field trips, like the sardines or peanut butter I brought when I was a student with little money.

I often incorporated field trips into attendance at meetings. This time, the annual meeting of the Texas Nature Conservancy was at Honey Creek Preserve. The members discussed ways to improve public relations and evaluated master plans for each of their preserves. I had taken over as the

Copperhead in a funnel trap

manager of Sheff's Wood near Tyler, which was the very first Nature Conservancy preserve in Texas. I received a small grant from the organization to set up drift fence traps for the long-term monitoring of snakes there. I had never used drift fences to catch snakes but knew about those used by Henri Fitch at the University of Kansas Field Station. These structures were solid metal fences about 50 feet long and were buried in the ground. The soil at Sheff's Wood was rocky, and I used a lot of volunteer student labor to dig trenches for the fences. The snakes encountering a fence would crawl along it and into a hardware-cloth funnel trap at the end. I checked the traps every other day during the warm months for 14 years, collected any snakes caught and individually marked them. The first few years, I marked them by clipping a series of scales. But in 1989, I started implanting microchips called passive integrated transponders (PIT tags). Texas Parks and Wildlife loaned me a PIT tag reader that they used with tagged fish. With the snakes, these tags were more permanent than the scale-clipping method. Although I learned a lot about snakes from this work at Sheff's Wood and published a few papers using data from it, the site was never used by my graduate students for their thesis work. They did not have the amount of time needed for a field project there.

When I took students to Cotulla, we always stayed at the KOA camp where Craig McIntyre had first taken me. At night, we drove the nearly deserted highways east of town. On one night, Carolyn and I found a really huge diamondback, probably close to 6 feet long. I had to put my whole hand around its neck to hold it while Carolyn took a photo with my camera. As luck would have it, the flash apparently did not work. We had no proof that I caught this massive rattlesnake! In those days, you didn't know how your pictures turned out until you developed the film. The rest of the night on this trip, we got lots of diamondback rattlesnakes but no checkered garter snakes. Carolyn really wanted to pin one of the diamondbacks. I wasn't very comfortable with her picking it up, but she was an adult, so she did catch one. I never was very comfortable teaching my students how to handle venomous animals, but at least I could show them the safest way. Actually, my second graduate student taught me the safest way to handle venomous snakes.

My second grad student was Jan Perry-Richardson, who came to work with me from the Fort Worth Zoo. We had known each other as students at the University of Kansas. Jan wanted to get a master's degree since she was passed up for a promotion at the zoo. It was not the first time for her, as it was common for males in Reptile houses of zoos to be given an advantage. Jan felt if she got better credentials, it might help her move up the ranks. She took a number of trips to Cotulla with me, and I learned much more from her than the other way around. Jan gave me insight into some of the politics and reality of zoo work. She also introduced me to a number of reptile curators. Probably the most valuable lesson I learned from her was how to tube venomous snakes. Tubing involved getting a snake to crawl into a clear tube that was just a little larger around than the snake. The snake couldn't turn around in the tube. And if we held its body and the tube, this prevented it from backing out. We could measure it and sex it very safely, since the snake's head was visible through the clear tube. Although it took a little practice, I quickly switched away from the pinning of snakes.

Jan did a really interesting research project with the checkered garter snakes. She studied how the females picked the best male of the many that would court them. From an elevated scaffold in my lab room, she videotaped garter snake courtship in large, 9-foot-wide cattle tanks. The females would try to get away from the courting males. As it turned out, the male that was best at keeping up with the female as she quickly crawled away was the one she chose. Apparently, the most energetic male had the highest fitness. After completing her degree, Jan did return to zoo work but had to work her way back up the ladder. Few women worked with snakes and the good-ol'-boy network was strong in herpetology.

Out on the deserted roads around Cotulla at night, we often saw interesting or unusual occurrences. Craig called them "little dramas." With Jan one night, the drama occurred when we both saw a snake crossing the road in front of us. As I slammed on the brakes and jumped out of the car, I yelled that it was a Mexican milk snake. Milk snakes are coral snake mimics, and as I ran up to grab it, Jan yelled, "No, it's a coral snake." I slowed down just a bit but yelled back, "No, it's a milk snake." She still argued, until I pointed my flashlight at the milk snake right in front of me. Then she pointed at a coral snake that was trailing right behind the milk snake. It turns out that coral snakes eat other snakes. The milk snake was going to be "dinner."

Another "drama" my wife Joan remembers was when we ran across a barred owl on the side of the road. I taught my ornithology students how to stuff the birds we found dead on the road. I had learned how back in Mr. Bee's museum techniques class at KU. So we stopped to bring the owl back to school. I picked up the owl and put it in a black garbage bag. The back seat was full, so we put the bag on the passenger-side floor boards, under Joan's feet. A few miles down the road, the owl woke up from an apparent concussion, extended a foot with sharp talons through the plastic bag and reached over to the gas pedal. I managed to pull over to the side of the road and let the wide-awake bird out of the car.

Barn owls often followed alongside our vehicle as we drove the roads of Cotulla. This species of owl seemed ghost-like, appearing white in the headlights and flying silently. I think they used the lights of the car to see mice in the grassy berms of the road. We stopped one night for one that was in the middle of the road. We thought it might be hurt since it didn't fly away. As it turned out, the owl was more interested in the garter snake that was trying to eat a toad that was stuck in the road tar. Another drama playing out.

I also took undergraduate students to Cotulla. One trip down to South Texas was with Fred Kersh, who was working with me on an independent project. Fred, like many of my other undergraduate volunteers, became an ER doctor. Maybe it was the adrenaline rush of both chasing snakes and handling medical emergencies! Fred and I were almost to Cotulla on a Friday evening when the gears in my Dodge Colt froze up. We managed to get it into first gear by pushing it with the doors open and then jumping in. However, we were in the small town of Devine, Texas, and all the garages there were closed. We debated whether to stay overnight at a motel but decided to try to make it back to Tyler, since the garages were probably closed for the entire weekend. Fred pushed the Colt again with the door open, I got the car into fourth gear and he jumped in. We drove all the way back — eight hours — without stopping at any red lights. We were lucky it was so late at night. We had to get gas once, but managed to find a station on a hill so I could get up speed before letting the clutch out. Students today most likely do not know how to "pop a clutch." Fred probably also remembers a trip we took to the Texas coast, when the tide came in and flooded our tent as we slept. But that is another story.

My Dodge Colt down at Cotulla.

Camping with a dozen or so junior and

senior undergraduate students was a very different experience than with one or two graduate students. I had the students plan the meals, and most were surprisingly good. Like the venison chili we had one time, or the barbecued quail and French fries another time. The meals didn't always work, like the time the students didn't cook the spaghetti noodles long enough. They were super "al dente," so we ate the sauce on English muffins instead. It actually wasn't too bad. The university didn't have vans, so we generally carpooled on trips and would sometimes lose a car or two. How did we survive before cell phones? The students had to use maps and plan ahead!

On a typical class trip, we set up camp and put out live traps for small mammals. We would then go out and drive the roads east of Cotulla for snakes. There were no real towns in that direction for almost 100 miles, so there was almost no traffic at night. We would stop in Los Angeles, Texas, a "town" in name only. It had one bar, which did have a bathroom. I would stop there around 2 in the morning so the students could use the facilities. The students would always get lots of stares. On those night drives, we always found lots of interesting species — Texas longnose snakes, *Rhinocheilus lecontei,* coral snakes, *Micrurus tener,* Mexican milk snakes, *Lampropeltis triangulum,* and big tarantulas and centipedes. We would return to our campsite around 3 a.m., so most students slept in until the sun made the tents too hot. The next morning, after scrambled eggs, sausage and bacon, we drove around the area looking for southern species of birds, like vermillion flycatcher, black-throated sparrow, great crested flycatcher and pyrrhuloxia.

On fall trips, we usually saw large flocks of Swainson's hawks migrating through. We would then drive around the same roads looking for Texas tortoise and indigo snakes. Most of the tortoises we saw were run over. I guess because they were slow. In all the times I went to Cotulla, we only saw one indigo snake and it was dead. Some years, we brought seines and recorded the fish we found in the river next to the KOA camp. Because it was always hot, we rested during the afternoon at the pool by the KOA office. The next night, we would go road driving again. The students liked to ride with me because I could turn my little Dodge Colt around on a dime when I saw a snake. We were stopped by Texas Parks and Wildlife just once — and they only wanted to see our snakes, not my collecting permit.

I took some of my trips to Cotulla with "amateur" herpetologists. These folks not only helped me get checkered garter snakes, they taught me many things about road driving, such as which roads produced the most, what weather conditions were best, etc. We often took multiple cars and split up, meeting somewhere later to exchange stories about what we saw. Usually, any car you saw driving slowly on these roads in the middle of the night would be other "herpers."

On one trip, I stopped in San Antonio to see Tom Vermersch, an amateur herper who made a living giving talks to science classes at local schools. He took me out to a site where he had placed some boards out. No checkered garters, but I got my first Texas patchnose snake, *Salvadora*

Trip with Alan Tennant and a Dallas reporter.

grahamiae. I knew trash piles were good places to find snakes, but this was the second time I saw someone deliberately putting out plywood boards. The other was in Chicago. In my future survey work, I switched from using drift fences to coverboards. Tom spread the word in the San Antonio Herpetology Society that I needed checkered garter snakes. The next time I came, Joe Laslo, curator at the San Antonio Zoo, and Mike Bishop from the club had lots for me.

In another notebook entry, I describe my trip to Austin to represent my department at the UT System committee meeting on animal welfare. I stayed with Craig McIntyre and went out hunting at another trash dump with him and Alan Tennett. We got patchnose snakes there; but at an old house east of Austin, we did get three checkered garter snakes. The next morning, Alan asked me questions about snakes for the book he was working on. It came out a couple years later as "The Snakes of Texas" (Tennant and Werler, 1982). To promote his book, Alan later arranged a trip we took to Cotulla with a reporter. Steve Smith wrote a fun story in the Dallas newspaper's Parade magazine about our snake hunting adventure.

Over the years, I took a variety of people to Cotulla to share how wonderful the valley of Texas was and to have company on my trips, but as it turned out, I also learned a lot from my companions. They helped me to slow down and enjoy the camping and cooking experiences. I also saw how different my students were outside the classroom. Students who did not do A work in my class often excelled on these field trips. Afterward, we often had different relationships in class. But more than anything, I learned how people who loved catching snakes came in all shapes, sizes and backgrounds.

I found that getting the students out of their comfort zone helped to engage them in the material.

CHAPTER 6

Adversity and Student Engagement: Costa Rica and Hunting Freshwater Mussels

I took biology students to Costa Rica for eight summers in the late 1980s. Even though we went in the dry season, it rained every day as we hiked up mountains to cloud forests. In later years (late 2000s), in an ecology class during the fall term, I took students to a local river to study aggregations of freshwater mussels. We took canoes to the mussel bed, and often, the students would tip them over. At the site, they had to go into the shallow water and dig up the mussels to count. On field trips in both these classes, I found that getting the students out of their comfort zone helped to engage them in the material. Our biology students were mostly premed majors, highly motivated to get good grades but less interested in outdoor experiences. Ecology was required for their degree, so I had the opportunity to take them outside. When years later, as doctors, they would come back to visit the department, I would ask what they remembered about their classes. They often recalled getting wet and muddy on field trips.

I began teaching the field class to Costa Rica by accident. My wife and I had been wanting to go somewhere in the tropics for a vacation over a spring break. Most biologists considered Costa Rica the best place to visit to see tropical forests. It had national parks in a variety of tropical habitats and good roads to get to them. We didn't have much money for a trip, so

I talked to Dr. Marty Crump, who had worked with amphibians in Costa Rica. I knew Marty from my time at the University of Kansas. She was at this time a professor at the University of Florida. I asked her what could we do over eight days for little expense. She sent me a letter with a plan of places we should go to and contacts for cheap accommodations. We followed her suggestions pretty closely in organizing our trip.

Joan and I flew into San Jose, Costa Rica, on March 15, 1985. We then took a harrowing taxi ride to Hotel Galileo, a student hostel ($16 a night for both of us) in the downtown area. Similar to what I had experienced in Haiti, drivers did not stop at stop signs. They zipped in and out of traffic and honked constantly. The next morning, we went back to the airport to rent a jeep so we wouldn't have to drive the crazy streets in San Jose. From the airport, we were able to directly take the easier drive up Highway 1 to Hacienda La Pacifica in Canas. Marty had said La Pacifica would be a good place to see the tropical dry forest in the province of Guanacaste. This low-elevation habitat had seasonal rainfall and was easily converted to pasture for cattle, so very little remained in natural condition. Guanacaste had the largest remaining dry forest in Central America, mainly as a result of the actions of the biologist, Dan Janzen, who pushed hard for its protection.

Howler monkey.

La Pacifica had a number of cabins with metal roofs on which large Tenasaur lizards would sun. Our cabin had a huge strangler fig in front of it. Strangler figs grow around another tree for support and eventually kill the tree, so the fig is hollow in its center. Each morning, a troup of howler monkeys woke us up early, and we would bird watch before breakfast.

We saw lots of species of birds, but because there were so many alike, we had trouble identifying them from our bird books. (These were Mexico and Central America field guides, as there wasn't one for Costa Rica yet.) The understory trees were less thick in this habitat, so we did get some good photos.

La Pacifica had a communal dining area, so we got to meet other interesting people and eat food we weren't familiar with. There really wasn't a specific Costa Rican cuisine, but the meals here were European in flavor since the owners were from Switzerland. Because of the pacifist nature of the country, many refugees in world wars migrated to Costa Rica.

Strangler fig.

In the next cabin over from ours, we met John Emlen, a mammal behaviorist from the University of Wisconsin. John and his wife helped us identify some of the birds we were having trouble with. They also warned us about the scorpions that were common at Monteverde Cloud Forest, where we were going next. This advice turned out to be fortuitous.

From Guanacaste, Joan and I headed up to see the Monteverde Cloud Forest Reserve. We stayed in the Pension Quetzal (Bob Law, owner) in the town of Santa Elena. We hiked up to the reserve the next two days. The trail in Monteverde was steep and muddy with steps created from logs holding the mud back. It led up to the Continental Divide, where we could see both the Pacific Ocean and the Caribbean! Epiphytic plants like bromeliads, orchids and ferns covered over the trees so thick that we could not see the tree's bark. These plants get nutrients from the air and do not have to have roots in the soil. There were also vines that started in the soil but grew rapidly upwards to reach the small amount of sunlight filtering through the mass of tree leaves. We saw beautiful birds zipping across the trails but didn't get very good looks at them. We actually did better birding down closer to our pension, where we could look down on birds in the treetops below. We did get to see the bird our lodge was named for. The Quetzal had brilliant lime green feathers and a long tail.

During our first night at the pension, Joan woke up in the middle of the night shrieking! She had been stung on her hand by something that crawled under her pillow. We scrambled around trying to find the light cord in the dark and then searched for the culprit. We finally located the scorpion in the bed sheets, and I used my boot to smash it! Luckily, we knew from what the Emlens told us that, even though the scorpion sting hurt like crazy, it wasn't severely venomous and no medical treatment was necessary. It was tough for Joan to get any sleep that night! A second shock occurred when we checked out. The bill was twice what we expected. We found out that hotel charges in most places were per person, not per room. At this rate, we would run out of cash!

After Monteverde we drove back to Hotel Galileo in San Jose to wait for our local flight to Quepos, which was scheduled very early the next morning. We took a taxi to the SANSA Airline office, where we were supposed to take a bus to the airport. However, the office was closed! We waited in the dark at the bus stop just outside the office, with a bunch of other people who were also taking the flight. A pay phone near the office rang and a local man answered it. We heard him talking in Spanish and managed to understand that "the plane was broke." We went back to the hotel, and luckily, they gave us our room back.

Later that morning, we decided to get a rental car and drive to Quepos. Actually, the canceled flight helped us, since our money was running low. The car rental office was one of the few places in Costa Rica that accepted credit cards, so we were able to save our cash. Joan and I drove the long road on the Pacific side of Costa Rica to Quepos, where we had rented a cabana at Manuel Antonio Park. The drive was boring, as oil palm tree groves lined the highway. I imagine this was why Marty recommended the plane flight. Our cabin at Manuel Antonio was small but nice, with a ceiling fan that ran sporadically through the night as the electricity cut in and out. In the park, we got to see several troups of Capuchin monkeys close up and lots more species of birds. Two days later, we drove back to San Jose and turned in the car.

From San Jose we took a bus up to the Poas Volcano by way of the town of Alejuela. The bus was crowded since lots of locals went to the park on Sundays. The bus went straight there without all the normal stops of a workday. On the bus, we met a couple who told us they were hiking throughout Central America. She funded their trip by belly dancing in bars.

Capuchin monkeys at Manuel Antonio.

We also met Jane, an entomologist from Leeds University. Lots of biologists would come to Costa Rica for vacations, since the country was the best place to see tropical habitats. The bus parked well below the volcano rim. We hiked the rest of the way up to look down at Laguna Botus, a lake in the volcano. This view was 2,650 meters above sea level and the high altitude gave me a bad headache. The park was in a cloud forest, and along the trail on the way back, we saw lots of bromeliads and orchids up in the trees. The moist air watered the plants.

The final day of our trip, we flew back to Houston but were very worried we wouldn't have enough money to get the car out of long-term parking! Luckily, we had received our refund from the "broke" plane flight in cash and had rented the jeep on the credit card. We had just what we needed to pay for our parking. Whew!

The next year, the university wanted faculty proposals to teach travel study courses to increase the global awareness of our students. Lynn Sherrod,

Neil and Bromeliads.

The Psoas Volcano.

a botanist in our department, and I received funding for a summer class that took students to Costa Rica. The first year, we designed a trip that mimicked the journey Marty Crump had created for Joan and me. Then, over the next nine years, we added different places that Lynn and I wanted to visit. We taught the course sometimes as field biology and sometimes as biogeography. We cross-listed the course for both undergraduates and graduates, so local school teachers could get credit. Although I wrote up each trip in my field notebook, I will describe just some of the adventures we had each time. Most of the students we took were from East Texas, and many had never traveled outside the country — or on a plane at all.

The class left from different cities over the years, sometimes Houston, sometimes Dallas and once we went through Miami. We often traveled on different Central American airlines (Taca, Sansa, Lacsa) and usually had multiple stops in Belize, El Salvador, Guatemala and Nicaragua. Sometimes, we had to deplane and wait in a local airport for refueling. The time we deplaned in Managua was a little scary. Nicaragua was having political unrest, and we did not know they were going to make us get off the plane.

On the first trip, we had 13 travelers and had allowed nonstudents to come along to help make the class. That was a strategic mistake. Some of the older folks on that trip expected a tourist vacation with nice hotels. In addition, a lot of our hikes were wet, muddy and slippery. That was OK for 21-year-old students, but it was worrisome when the older folks fell. The first night, we would stay in San Jose, at either Hotel Europa or the Gran Hotel. Not fancy enough for the old folks but fine for students. A tourist bus with our guide would pick us up the next morning. Our first stop usually was the Psoas Volcano. We had to get there early because it was a geyser volcano that would erupt and form steam, becoming impossible to see. On the trail up to the volcano, Lynn would explain about the epiphytes (bromeliads and orchids) living all over the trees. I would start teaching the students the birds here, as the trees were shorter and birds were easier to spot.

Generally, our second day in Costa Rica was to Selva Verde on the Caribbean side or Finca La Pacifica in Guanacaste. La Pacifica had lots of neotropical birds like toucans and tanagers, but the students really enjoyed seeing the howler monkey troup there. Richard Leval, our guide for most years, had studied bats for his Ph.D. at Texas A&M, so we brought mist nets (with fine threads the bats could not detect) and would set them up in front of the flowers of peperomia plants to capture nector-feeding bats. One night, we got five species of leaf-nose bats. Most of the species in the U.S. are insectivores but not in the tropics. On only one trip did we collect any vampire bats.

We always saw the most snakes in this dry habitat, including cat-eyed snakes and, one year, a coral snake. Even though most people think of the tropics as having lots of snakes, we actually saw very few. Of course, the exception was the year a woman with a severe snake phobia went with us. That year's trip was nicknamed by the students as the "Snakes Alive" trip.

Richard Leval catching bats in a mist net.

Even then, we only found 16 snakes in the two weeks. However, almost every snake we saw that year was a different species. That high diversity was also typical of other animals and plants in the tropics. For example, there were five species of kingfisher here, whereas in Tyler, we only had a single species. At La Pacifica, the howler monkeys would wake the students early in the mornings. I was once awakened by a lesser anteater *(Tamandua)* as it scratched open a termite nest in a tree above my cabin. As the sun warmed the large tenasaur lizards, they also made lots of noise on the metal roofs of our cabins. Geckos would squeak through the night in our cabins, but as the students said, "sleeping wasn't that important on these trips."

We often took the students to see the animal rescue facility next to La Pacifica. They had a number of native cats that could not be returned to the wild. One year, as I was getting off the bus, I looked up and a student, Ken Hovey, had blood streaming down his face. Ken, who had lost his contacts the day before, moved in a little too close to see a caged jaguar. The cat reached through a hole in the wire cage and struck Ken in the face. A claw caught him above his eye, but he pulled back fast enough that it only grazed his cheek. He was so lucky! We had a podiatrist on that trip, and he put a butterfly clamp on it. On that same trip, another student petted an ocelot that was purring in its cage. It bit her finger and tore her fingernail off completely. While bandaging the student's finger, the podiatrist just said, "Fingers, toes, they are all digits anyway." The rest of that trip, I kept telling students to back away from things!

From La Pacifica we would go up to Monteverde, the highlight of each trip. It was a cloud forest preserve near the town of Santa Elena, founded by Quakers during the Korean war. After bailing out their sons who were jailed for not fighting, several Quaker families moved to Costa Rica because it did not have an army. The families bought land on a mountaintop, raised milk cows and started a cheese factory. At night, we hunted glass frogs in the stream below that factory. The frogs would stick their eggs on the underside of big leaves, since egg masses laid in the fast-flowing water would wash away. When the eggs hatched, the tadpoles would drop into the stream below. The frogs were called glass frogs because their heart and blood vessels could be seen through their clear belly.

During the day, we would hike up to the preserve on wet, muddy trails to the Continental Divide. In the afternoon we would walk in the lower forest, where we would see and hear the three-wattled bellbird (the males made a tremendous bong noise) and sometimes the resplendent quetzal (the national bird of Guatemala). The first year we went there, we saw a Golden Toad in the nature center at the entrance to the preserve. The next year, they were extinct. The global decline in amphibians from chytrid fungus

Hunting glass frogs.

had wiped them out. Although we didn't know it at the time, they would never be seen again.

Hiking Monteverde.

We generally went to the Pacific Ocean side of Costa Rica for a couple days. There were several nature preserves there, including Manuel Antonio and Corcovado. It was a long drive, so we took small planes instead. The year we went to Tiskita Lodge, the planes landed on a narrow strip of beach. The pilot first buzzed over the landing strip to scare the cows away. The one time we did drive down the Pacific coast was memorable, because we stopped at the river in Caraca National Park. We saw flocks of scarlet macaws there, and at the bridge, we all watched a large American crocodile sunning on the bank. Then someone spotted a 6-foot boa on the steep bank. I slid down and grabbed it but couldn't get back up the steep bank with it, as it was biting my shoulder. Several students reached down and pulled me up. Not sure it was worth the trouble. But the students thought it exciting, as the croc was waiting for me down below.

The students loved Manuel Antonio Park because the troups of capuchin monkeys would come close to us. On another trip, we went to Drake's Bay Wilderness Camp further down on the peninsula of Corcovado. We took several small boats through the choppy water to get there. I got seasick when we stopped to look at the blue-footed boobies on an island (gas fumes from the boat didn't help). I ended up with a bad headache and missed out on some of the great food! We took several long hikes.

I remember having to push a very overweight student up a muddy hill. Then she immediately gave up on the hike and I had to help her back too! On that same hike, we ran into killer bees in a dead tree on the trail. We had to sneak by it. The guide said to run if they attacked! One of the students waited until we safely passed the tree to reveal she was allergic to bees and had forgotten her EpiPen. Our guide told us a student had recently died from killer bees in Costa Rica. The student had fallen and caught his foot, and his friends could not get to him because of the bees. What a story to tell as we were hearing them buzz!

There were some really huge trees on this trail. They extended above the canopy and had buttress roots, like wooden planks, to support them in the shallow tropical soils. We did see hognose vipers on the trail occasionally. One clueless student stepped on a coiled-up juvenile and then asked, "Who put that there!" One night, the students caught a tree snake, *Imantodes,* and I explained how these extremely skinny snakes ate glass frog eggs. The eggs are sticky and hard for the snake to swallow. So when the snake grabs the eggs, the tadpoles detect the predator, begin hatching early and fall

Tree with buttress roots.

A tree snake *(Imantodes)*.

into the stream below. The next day we snorkeled in a cove in the bay and saw sergeant majors, wrasses and other beautiful fish. The water helped the backache I had from pushing and pulling the overweight student.

We didn't always take trips to the Caribbean side of Costa Rica, as it was a long drive. The first year, we did go to Cahuita National Park, but it was very small and the coral reef there was dead. I was swimming in a pool between some coral heads when an 8-foot shark appeared. I got out quickly but cut my heel on the coral and started bleeding. Oh darn!

We stayed that night in Limon, which was the poorest city we saw in Costa Rica. The Caribbean side of Costa Rica had been involved in the slave trade and was predominantly Afro-Caribbean in culture. Our rooms were also pretty poor, so one trip there was enough. The next year, we took a day trip to La Selva, the Organization for Tropical Research's station. We couldn't afford to stay there overnight, so we lodged at Finca Selva Verde, which was beautiful. I loved watching the cooks make coffee in the morning. They dumped a whole pound of coffee in a big pot, boiled it and then scooped the coffee off the top — mmm! We engaged in some of

Finca Selva Verde.

the best birding at La Selva. There were trogons, woodcreepers, motmots, oropendolas, manakins and dozens of other species. We saw a slaty-breasted tinamous, and I explained to the students how two brothers in this species share one wife. It takes all three to raise the young, and the brothers receive some genetic benefit from raising each other's chicks. On the trails (with boardwalks), we saw several snakes and I couldn't help but catch them. I did get scolded by the guide, as everything was protected here.

We stopped at the Barillo Carillo Preserve on the way back to La Pacifica and walked a trail there. We found a very large boa in a hollow tree but couldn't get it out. We saw two kinkajous talking seriously to each other — and wondered if they were having a fight. Barillo Carillo also had the only spider monkeys we saw on any of the trips. We once stopped at the Arenal Volcano, which was releasing plumes of smoke. Our bus got stuck in the

Swimming pool heated by Volcano Arenal.

surrounding ash field, and all the students had to help push it out. We also stopped at a swimming pool that was heated by the volcano.

We always ended the trip with a day in San Jose so everyone could do some "tourist" shopping. We usually had a party at our house a few weeks later, and it was always the more "dangerous" events that the students remembered as being significant to them. From then on, as often as I could in my classes, I would get the students out in the field and uncomfortable.

Advances in science occur much quicker when researchers with different skills work together.

CHAPTER 7

Collaboration: A Summer at the Savannah River Ecology Lab

One of the most important characteristics that I associate with my success in conducting field research is my collaboration with colleagues. Advances in science occur much quicker when researchers with different skills work together to solve problems or answer scientific questions. In fact, it is often required. Most of my projects started as discussions with colleagues about observations we had made. From there, we developed ideas to figure out why some particular activity occurred. I think this verbal interaction is common in field biologists, particularly because they often wonder about certain activities in the animals they study. Other biologists may see the same behavior in their subjects and generate ideas as to why this might be occurring. Conversations about these activities often lead to collaborations to test their hypotheses. I also just enjoyed those discussions!

I met Dr. Richard Seigel at a scientific meeting in Salt Lake City in 1984, when we both gave presentations about the large amount of variability in size that we saw in newborn baby garter snakes. These differences in offspring sizes occurred even when the pregnant females came from the same location in the same year. We both believed the variation in offspring size was related to the amount of foraging success the mothers had. After some discussion during breaks between talks, we decided it would be easy to test in the laboratory with the checkered garter snakes I collected from Cotulla. Rich had just started a postdoctoral position at the Savannah River Ecology Laboratory (SREL) in Aiken, South Carolina, and he invited me to bring my animals to SREL the next summer to study the effect of energy intake on snake reproduction.

With Rich's help, I received a Faculty Participation Grant from the Oak Ridge National Laboratory to work at SREL in the summer of 1985. This facility was the premier ecology laboratory in the U. S. for long-term studies of reptiles and amphibians. It was established 70 years ago, when a large area of land was allocated for the government facilities where plutonium was produced for nuclear bombs. In conjunction with that work, the University of Georgia set up SREL to conduct research on the protected land.

Rich and I had very different personalities and brought different strengths to the project at SREL, but we found we worked very well together. Over the next two decades, we collaborated on a number of research projects examining the role of energy intake on the reproductive characteristics of snakes.

Rich at SREL.

I brought my colony of checkered garter snakes with me to Rich's laboratory at SREL, and we started our first project studying phenotypic plasticity of reproduction. Phenotypic plasticity refers to the ability to change a trait such as litter size, based on some environmental variation such as foraging success. Our project at SREL was simply a test of the effect of dietary intake on the number and size of babies these snakes produced. We hypothesized that female snakes with a higher energy intake would have more or bigger babies. Our high-diet female checkered garter snakes were fed three times the amount of food as the low-diet snakes. Snakes were particularly good models for this research since they could reproduce even on the reduced-energy diet. This collaboration ended up as a publication in Ecology, a high-impact journal. By the way, the high-diet females had more babies, but their babies were the same size in length and weight as those produced by low-diet females. So we did not answer our question as to why wild-caught females had such variation in offspring size.

Rich and I did several other lab studies at SREL with the checkered garter snake mothers and their babies. One paper that has been cited a lot was conducted by an undergraduate student, Mary Huggins, who was on an internship for the summer. Mary examined the locomotor capabilities of the pregnant snakes. She chased the female garter snakes with a broom around a circular race track. She found that females carrying more babies were slower and tired quicker. Not rocket science, but the results were important as they suggested that slow pregnant females would be subject to higher rates of predation. Females may need to limit the number of offspring so they can escape predators. This may seem obvious, but very few tests of this idea had been conducted. It also might explain why pregnant female snakes were known to quit foraging. Not because they won't eat, but because it is too dangerous to go out looking for prey.

SREL was established in the 1950s on a remote area of South Carolina that was set aside for four nuclear reactors producing plutonium (an additional one was shut down). When the reactors were built, the Department of Energy moved everyone off the area to have lots of room in case of an accident. Through an association with the University of Georgia, biologists began long-term ecology studies there since the land would be relatively undisturbed by humans. The Department of Energy supported the

Carolina Bay at SREL. (Photo by Rich Seigel)

work, and the facility began to attract a number of visiting professors and graduate students to work there. The director, Dr. Whit Gibbons, and his students were particularly interested in studying the reptiles and amphibians living in the Carolina Bays located on the property. The Carolina Bays were ponds that formed from meteor strikes. The bottoms of these elliptical pools were all deeper at the same end since the meteors hit at an angle. The ponds had large numbers of breeding amphibians, which in turn supported high populations of aquatic snakes that lived on the tadpoles and salamander larva there. I particularly liked the black swamp snakes, *Seminatrix pygaea,* that reminded me of that first worm snake I caught as a child.

There were also creeks and lakes on the property, where the hot water released from cooling the reactors made them shadowed in steam. That hot water killed the trees, and the streams looked very eerie. We would hear frogs calling in other streams at night but not in those heated ones. Only alligators seemed to be able to get out of the hot water quickly enough to live there. Besides getting animals in the Carolina Bays, we drove the roads at night looking for snakes and had to avoid tanker trucks taking away the waste

Stream during hot water release and afterward at SREL.

reactor material. These 16-wheelers were always going superfast to limit radiation exposure to the drivers.

I also helped seine for turtles in ponds in areas off SREL property but nearby the facility. This project was funded to determine if turtles might have left the property after growing up in SREL pools that were used to store low-level radioactive waste. People in South Carolina did occasionally eat locally-caught turtles. If the turtles were from SREL, they had some radioactivity. I remember a team of us pulling a huge net across one pond below a pig farm. The pig waste had washed down into the pool, filling it with mud and feces. The few turtles we got there had deformed shells.

One day, when I was sitting in my lab, someone from the turtle crew walked in with a burlap sack full of turtles. I watched him scan each turtle with a Geiger counter and then heard the machine go buzzzzzz. I asked him if it was dangerous to be sitting near them and he said no, it was just low-level beta radiation. Coincidentally, when we started at the lab, everyone had to take a thyroid gland test. The results always came back as indicating a hyperactive thyroid. The thyroid is apparently the first organ to show

Seining for turtles on a pig farm near SREL.

effects of radiation. Humm! My wife, Joan, came to visit in July, and as she was pregnant with David, we were a little concerned. We decided to visit Charleston and not spend any time at the lab.

I also helped the "gator crew" a few times that summer. We once were asked to track an 8-foot-long alligator, which had a radio collar, in a slough on the property. Although we knew from the rapid beeping of the radio that the alligator was in the shallow water we were walking through, it was amazingly hard to find. We would walk up the creek, where the beeping radio told us it was. And then, all of a sudden, the radio would start beeping behind us!

I also went with the crew on an alligator removal trip at the Kiawah Island golf course and housing area. People there would feed the alligators from their back porches, but would be upset when a big one came into their yard! On that trip, we caught one that was 11 feet long. The crew strapped it to an extension ladder and put it in the large van we were in. I rode in the back with it — and learned that alligators are really strong! As it thrashed in the van, I moved as far away from it as possible.

The SREL was a successful lab not only because of the lack of human disturbance on much of its property, but also because of opportunities for interaction between large numbers of scientists. Collaborations often sped up research because of the varied backgrounds and abilities of those colleagues. I have worked successfully with a number of collaborators over the years, but I worked best with Rich. At SREL, we would discuss our research throughout the day and night, coming up with questions we wanted to explore. Then we would develop a plan of action. Who would conduct which experiments? Who would write what part of the paper? We took turns serving as project lead, knowing the next project would be led by the other person. I published four papers based on our work that summer. We often let students conduct parts of the research, knowing they would need publications to get future job offers. All my interactions at SREL were enjoyable, and the excitement of interacting with other scientists made the work seem less tedious. Even the security badges we were required to wear were tolerable.

This grant allowed us to compare reproductive traits in checkered garter snakes from Texas and Arizona.

CHAPTER 8

Funding for Fieldwork: A Grant to Work on Reproductive Variation in Garter Snakes

Although field biology research can be conducted on limited funding, grants are often needed to pay students and travel expenses. I had very limited funds for my work at UT Tyler on pheromone trailing in snakes. I received a couple of small, in-house grants and used them to build an arena to videotape trailing behavior. These grants also supported my trips to Cotulla to collect checkered garter snakes and to pay salary to undergraduate students to run some of the time-consuming experiments. However, in 1988, Richard Seigel obtained a larger grant from the state of Louisiana for us to take several trips to Safford, Arizona, to collect the same species of garter snake used in our SREL study on reproduction. This grant allowed us to compare reproductive traits in checkered garter snakes from Texas and Arizona.

I joined Rich and his students on trips to Stafford in October of 1988 and again in 1989. I met him in Houston and left my car at the Houston Zoo. I went by myself on the first trip but brought undergraduate students along on the second trip. It was a long drive in his university van, so Rich played a game with the students. They had to guess the species of amphibian or reptile when he read the description from a field guide. Rich was serious about training his graduate students to be herpetologists, and it was good for my students to see that dedication.

Roper Lake, Arizona.

Rich with students (Sean Doody, Tom Blanchard, Rich and Bill Lutterschmidt).

In Arizona, we stayed at Roper Lake State Park, just south of Stafford, and collected garter snakes at that lake. But most came from a subunit of the park, Dankworth Pond, located, 3 miles south of Roper Lake. Because of the arid habitat, the garter snakes in both places were restricted to the water bodies. Those checkered garter snakes I had caught in South Texas were less restricted in the habitat where we found them, and we hunted them on roads at night. In Arizona, we used minnow traps on Dankworth Pond and in an overflow stream of the pond. The snakes fed on bullfrog tadpoles, and those amphibian larva would get caught in the minnow traps and then attract the snakes. Interestingly, bullfrogs were not native to Arizona. They were brought in with the mistaken idea that bullfrogs could be harvested. Instead, they were a serious problem, as they fed on native wildlife, even baby garter snakes.

Rich checking minnow traps.

One interesting side trip we took was to Mt. Graham, located nearby. Since it was biologically isolated, it had some endemic species such as the Mt. Graham red squirrel. We did see the squirrel but wanted to find the terrestrial garter snakes, *Thammophis elegano,* that lived there. We didn't have any luck as it was surprisingly cold. It was amazing to get up to 10,000 feet, after being in the desert just a few hours before.

We brought back adult female checkered garter snakes to my lab and got reproductive data when they had their babies. Then we artificially hibernated them. When we brought them out, we set them up in two groups receiving different amounts of food. Then we collected data on the babies they produced. This was the same experiment we had conducted at SREL on animals from South Texas, now applied to the females from Arizona. The females from Arizona had much larger babies than those from South Texas, but interestingly, the Arizona females' diet did not affect the number or size of their offspring. The South Texas females with more food had more babies than the females with less food. This change in a trait with environmental differences is called the "norm of reaction." We hypothesized that the snakes in South Texas had higher norms of reaction, allowing them to respond to the year-to-year variation in rainfall that occurred there. In contrast, the snakes in Stafford, Arizona, lived around ponds that produced a more stable food supply.

Getting grant funding for me has always been a mixed bag. Obviously, money is helpful, particularly in attracting good graduate students and helping undergraduates with their education. Paying them a salary has become critical as tuition costs have increased. However, writing grant proposals takes away a lot of research time, and the success rate is generally pretty low. I would guess I wrote about six proposals on average for each grant I received. Then, annual and final reports that are required take as much time as writing up the research for a publication. Luckily on this grant, Rich did a lot of the reports. Actually, many colleagues on my grants have helped with some of these less enjoyable parts of research funding. Another good reason to collaborate!

Another way I was able to fund students was by conducting surveys for conservation projects. I was hired to survey the reptiles and amphibians, the birds and the small mammals of a Wildlife Management area called the Old Sabine Bottom (see Chapter 14). I used the funds to support graduate students while they conducted their own research. Generally, the research was applicable to conservation, as with Paul Hampton who studied the impact of the food plots that Texas Parks and Wildlife created on old oil field pads in the management area. Tim Hunkapillar studied the effect of the ATVs traveling on the trails there while he was gathering data for me on the abundance of the various lizards, snakes and amphibians.

I did similar work for the Army National Guard at Camp Maxey near Paris, Texas. Again, the funds they paid for the report on the species of the area were used primarily as salary for the graduate students. Since they all enjoyed going out "herping," they would have worked for free, but it was great to be able to support them.

It was always helpful and sometimes critical to have the right equipment.

CHAPTER 9

Blisters, Bugs and Being Prepared: What I Learned Collecting Red-bellied Racers in the Caribbean

Each field trip I took reinforced to me the importance of being prepared with exactly what I might need for that voyage. From having to write data on the leg of a student because we had forgotten to bring the field notebook, to running out of food and water on a hike that turned out to be much longer than expected, to needing the proper tools to pick up venomous snakes, it was always helpful and sometimes critical to have the right equipment.

In March of 1997, I traveled to the islands of St. Eustatius and Saba to collect red-bellied racers to bring back to my lab at The University of Texas at Tyler. This project originated after I met Mark Day of Fauna and Flora International at the annual herpetology meeting in 1996. At that meeting, I learned about the European conservation organization's work to save the endangered Antiguan racer, *Alsophis antiguae,* on Great Bird Island in the Caribbean. This snake had been considered extinct since 1936. The mongoose had been released on many of the islands in the Caribbean to control rats. Instead, the rats had extirpated native animals including this snake. There were no mammalian predators on most of the Caribbean islands, so the snakes had not evolved antipredator strategies and were easy prey for the mongoose. The rats had evolved with mammalian predators and were much tougher prey for the mongoose. So the rat populations remained high, and unfortunately,

Dr. Jenny Daltrey on Great Bird Island.

the Antiguan racer was completely wiped out on the island of Antigua.

Amazingly, over 50 years later in 1989, one individual Antiguan racer was discovered on Great Bird Island, a very small island located just south of Antigua. The discovery of that specimen reached Fauna and Flora International. They hired Dr. Jenny Daltrey to study the snakes and begin efforts to increase their numbers on that island by removing the rats (https://www.fauna-flora.org/news/a-race-against-the-odds). However, since those snakes on Great Bird Island were the only known population of the species, the biologists knew the species could easily be wiped out, i.e. by a hurricane. In order to increase the likelihood of the snakes' survival, the conservation organization was interested in establishing the species on nearby islands. That's where I came in.

One method biologists use to repopulate species is to breed them in captivity and release them in habitats where they used to exist. It is a strange fact, but because there is usually more funding to study endangered species, we often know more about rare species than we do common species. However, because the Antiguan racer was so rare, the staff of Fauna and Flora hesitated to take any of the snakes from the wild. In my lab, I worked with common species that were easy to maintain and breed in order to learn about snake reproductive strategies in general. I told the conservation organization I would be willing to study reproduction on a related species of racer to provide better information for successfully breeding the Antiguan racer in captivity. This would allow them to remove fewer individuals from the wild.

Fauna and Flora International arranged for me to collect 10 red-bellied racers, *Alsophis rufiventris,* from the island of St. Eustatius to bring back to my lab at The University of Texas at Tyler. This species was known to still be common on both St. Eustatius and Saba because the mongoose was not on these islands yet. I flew to Antigua to meet up with Dr. Daltry, who was radiotracking Antiguan racers on Great Bird Island at the time. She was going to go with me to help collect the red-bellied racers. I was met in Antigua by a representative of their environmental awareness group. Thank goodness he was there at the airport with a sign saying "Niel" Ford, as I didn't have a phone number for them!

The next day we took a boat over to Great Bird Island to pick up Dr. Daltry. This island was uninhabited and Jenny had been there a week. She was very happy to see us, as she didn't have any way to communicate back to the mainland. Jenny was English, and before she joined this project, she had been working with King Cobras at the Madras Crocodile Bank in India. She showed me two Antiguan racers she had just caught. They were beautiful animals, the males brown in color and the females silvery, and both with

Great Bird Island, home of the last remaining Antiguan racers.

pink spots. This sexual dimorphism is very unusual in snakes, as males and females of most snake species are typically the same color. We went hiking around to locate an animal that had been implanted with a radio. Great Bird Island was very small and was covered in lava rocks, so walking was very tough. The vegetation on the island consisted of scrubby, desert-like plants. As we explored, Jenny told me of the time the rats got into her water supply and the boat was not due to come back for several days. Since there was no fresh water on the island, she had to drink cactus juice. She said she was just about to try to swim to Antigua (20 miles away), when the boat arrived. This was clear example to me of the challenges of working in other countries. Here however, after our hot hike, I got to swim in the little bay to cool off!

On Great Bird Island we saw lots of Ameiva (Pholidoscelis), a medium-sized, slender, fast, striped lizard. They were the primary prey for the snakes. Because these lizards had also evolved without mammalian predators, they appeared almost tame and even ate cheese right out of our hands. Jenny and I measured the snakes we had caught that morning and then released them back where we had collected them. Jenny packed up her gear, and we went down to the bay where the boat had returned to pick us up.

Male and female Antiguan racers.

The next day, we flew from Antigua on Liat Airline to St. Martins and then changed planes to Win Airlines for the short flight to the island of St. Eustatius. The plane we took was similar to the little American prop engines that flew out of my hometown in Texas. St. Eustatius, or Statia as it is called by most people, was a small Dutch island with only about 2,000 residents. The hotel we had planned to stay in was full, so we had to go to King's Well Hotel in Oranjestad, instead. It was more expensive, so we decided to share a room to save money. After checking in and dropping our gear, we walked up to an extinct volcano on the island called the Quill.

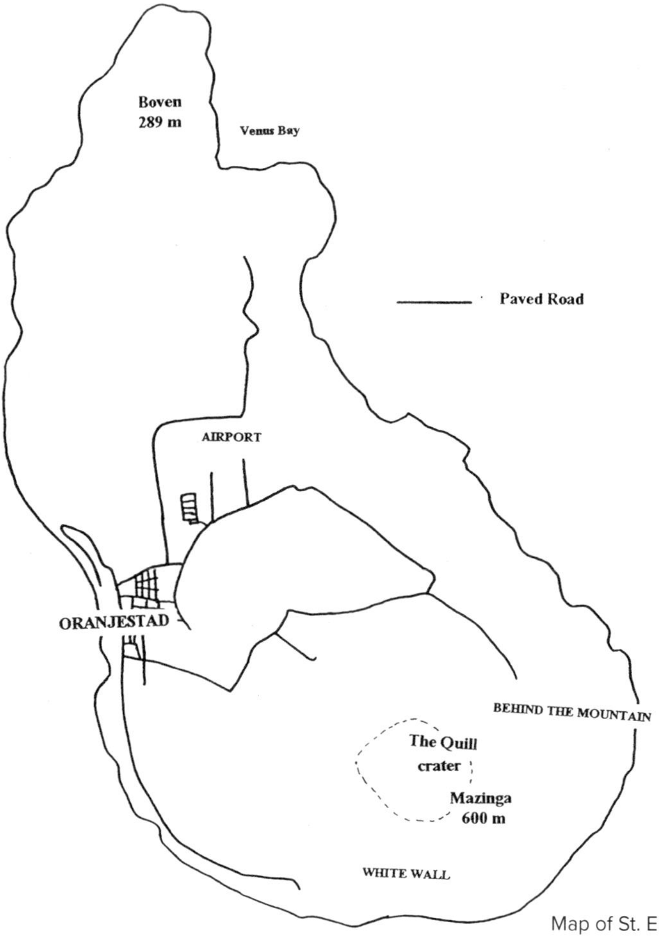

Map of St. Eustatius.

The Quill, an extinct volcano on St. Eustatius.

Jenny had learned that, because the center of the volcano was deep, the forest was too hard to log, so there were still lots of trees inside the crater. We followed a narrow dirt road called Rosemary Lane up to the rim of the volcano. The rocky slope into the volcano was treacherous. Hurricane Louis had hit the area in 1995 and blown down trees everywhere. In addition, there were numerous rock slides with large boulders. We had to climb up to the rim, clambering over the boulders, and then down into the basin of the volcano by holding onto the fallen trees. Both of us did kind of enjoy the physical challenge and competed to see who could get up and over boulders first. I lost! Jenny was used to scrambling over lava rocks on Great Bird Island, so she had a real advantage.

In the basin of the Quill, the ground was flatter and it was easier to walk around. However, the forest was thick, and we were a bit worried about getting lost. There were a few orange and pink flags marking the trails, but the paths themselves weren't very clear. We quickly caught several red-bellied racers, as they really didn't try to get away. Their only predators on St. Eustatius were hawks, so when we approached them, the snakes just crawled a short distance away and then stopped. However, as we chased the snakes, we realized we had lost track of the trail. Luckily, we could see one of the sloped sides of the volcano, so we headed in that direction and just started climbing up the boulders. Since this route was not a trail, it was an exceptionally difficult climb. We had to rest awhile after reaching the top. However, it was quite hot on the rim and we hadn't brought any water. There were also lots of annoying bugs, so we couldn't pause for very long. On the walk back to the hotel, there was a little store where we could buy drinks. Thank goodness we at least had some money with us. It was another lesson to me to be prepared with food, water and bug spray on any hike, no matter how short I thought the hike would be. We could have been stuck in the Quill for a while and our outcome even less pleasant.

Rosemary Lane, up to the Quill.

A Lesser Antillean Iguana.

The next day, Jenny had made arrangements with a local doctor to help us look for the Lesser Antillean Iguana, an endangered lizard. I decided to go back to the Quill by myself for a longer hunt for snakes. But on the long hike up there, I developed a bad blister on my heel. I switched out of my hiking boots to the tennis shoes I had in my backpack and left the boots behind a tree. The tennis shoes didn't provide as much traction on the rocks or ankle support as my boots, but I didn't really have a choice. I couldn't afford for the blister to get any worse. After making it over the rim and into the basin, I knew my way better this time and didn't get lost. I caught several snakes without much trouble, although there were hermit crabs crawling everywhere on the forest floor, sounding like snakes moving through the leaf litter. I hadn't brought enough bags to put the snakes in individually, and it was challenging to get the snakes bagged up. They tended to try to sneak out the top when I was putting another one in. I quickly caught all the animals I could handle and headed out of the basin, but I constantly slipped on the rocks in my tennis shoes. I finally got over the rim and started down the

Rosemary Lane trail. Then I remembered leaving my boots behind a tree inside the Quill. I had to hike back over the rocks once again! My calves and lower back really ached by the time I reached the hotel.

Back in the room, Jenny was measuring the snakes we had caught the day before. She did this while sitting on her bed. The animals were not happy and defecated all over the bed covers. She was too engrossed in what she was doing to notice, but the maid who cleaned our room the next morning may have wondered what happened on that bed. As we picked out some of the snakes to return to the Quill, we noticed most were males. In the two days we had been on St. Eustatius, we had already collected more male red-bellied racers than my permit allowed me to take back to the states. We would need to concentrate on catching only females the following day.

After apple pancakes for breakfast, we took three male snakes back to release. I had to walk slowly because of my blister, so Jenny went ahead. Going slower actually allowed me the time to take some pictures of a beautiful orchid on Rosemary Lane and also of the White Wall, a famous limestone formation on one side of the Quill. In the basin that time, we had our hunting technique down and got lots more snakes, but again, most of

The trail into the Quill.

them were males. On the way back to the town of Oranjestad, we stopped at the little gift shop again for fruit drinks. I guess we had developed a taste for them, as I had brought plenty of water that time.

The following day, we went with a park ranger, Ferdinand Von Staelandberg, up to the north end of the island to an area named the "Boven." Ferdinand told us that a national park was planned there and, indeed, it is there today (https://www.statiapark.org/). Up on a high ridge, we caught several iguanas, but as I grabbed for another one, I felt something sting my right calf. I slapped at it, and it immediately stung my left calf. It hurt like crazy! Ferdinand said it was a type of wasp called a "Jacky" and the swelling would last a week. We hiked up to the Quill that afternoon and my calves were swollen, so I was forced to walk at an even slower pace. Still, we caught more racers, some of which were females that time. Because the sexes were different colors, it was easy to concentrate on getting the silvery females. After we thought we had enough, Ferdinand dropped us back at the hotel. I was very glad not to have to walk any further with my aching calves.

At dinner that evening, Jan Faber of the Dutch National Park Service showed up to help us with the export permits. Jan was retired military, and he and Jenny spent the evening talking about survival techniques. If Jenny ever ran out of water on Great Bird Island again, she planned to try catching the morning dew with a tarp. The restaurant filled up with people from a cruise ship that had docked, and with so many folks talking at once, it was hard to hear Jan and Jenny. So I went back to the hotel room.

We still had several days before our departure, and we had caught all the animals that I could to take back to Tyler. Jenny decided we should look at the red-bellied racers on the nearby island of Saba. So that morning, we went up to the Quill one last time and released all the extra animals. The hotel manager drove us to the airport and provided us with the name of a good hotel on Saba. Although the locals on this island thought it strange that we were collecting snakes, they had all been very friendly and helpful.

The plane to Saba was another little eight-seater Saab and the flight was very short. As we prepared to land on Saba, Jenny told me that the runway was the shortest commercial runway in the world (400 meters). The island was just the top of a mountain coming out of the ocean, and there was only one little area of flat land on it. Everyone on the plane held their breath as we landed, and then clapped when our pilot managed to stop without crashing off the cliff.

Jim Johnson from the Saba Conservation Foundation picked us up and took us to the Hotel El Momo. The rooms were actually cabins built directly into the steep slopes. Even the pool was built into the mountainside. Saba was a popular tourist hiking paradise and there were several well-maintained trails up into the forest. Jenny hiked barefoot and I got by with my tennis shoes. The trails were narrow paths, however, with very steep banks on both sides.

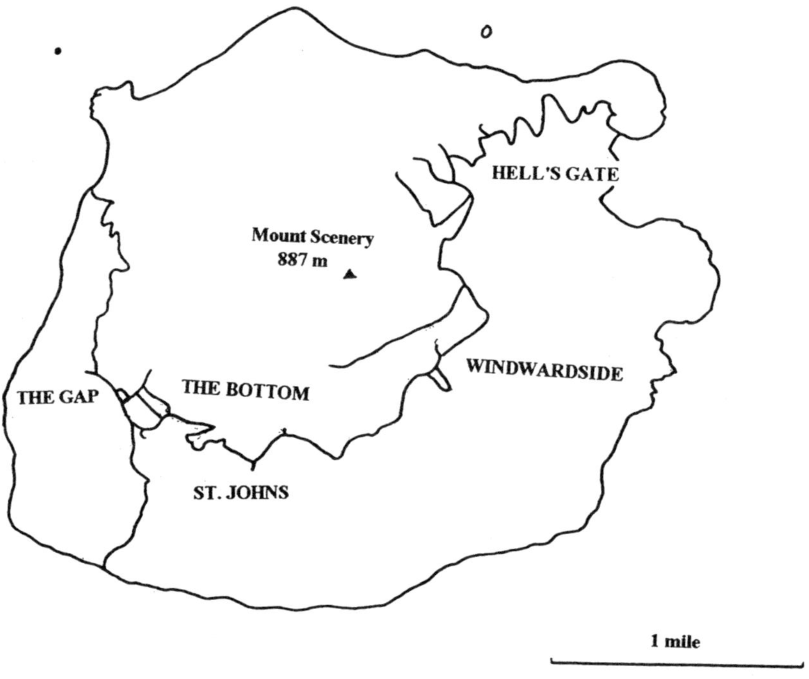

Map of Saba.

We missed the first two snakes, as they were more alert here than those in the Quill and they quickly dove over the steep slopes. Diving for the first one, I almost tumbled down the mountainside. Only the thick vegetation on the bank stopped my fall. I finally realized I had to look further ahead to see the snakes before they shot off the trail. I managed to catch two. We measured them and took photographs, but they looked identical to those on St. Eustatius. Jenny noted that they both had some head-scale anomalies, and she wondered if the snakes might be inbred on this small island. I also took some pictures of the Saba anole that we saw along the trail. It is an endemic lizard that only occurs on the island of Saba. It looked much like the green anole that was so common in the Southeastern United States.

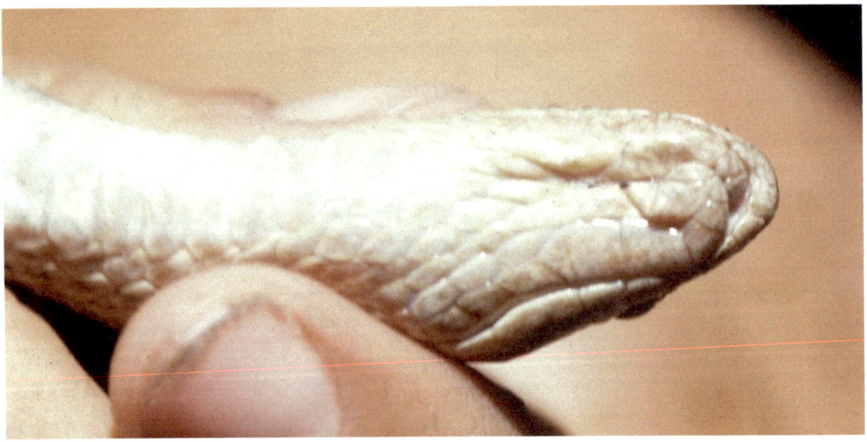

Scale anomalies on red-bellied racers from Saba.

That evening we splurged and ate grilled goat and curry shrimp at a fancy hotel. Afterward, Jim dropped us back at the Hotel Momo, where the owners, Els and Qed Rommers, had thoughtfully pushed the two twin beds together for us, thinking we were "together." We pushed them back apart. However, they had only left one blanket, so I got pretty cold that night.

The next morning we hiked a different trail named Sandy Cruz. The snakes were even harder to catch there, but I finally managed to grab a couple of them for us to look at. Tom VanHoff's house was at the end of this trail.

The Saba anole that is endemic to the island.

He was the head of the Saba Marine Park. He and Jenny discussed the endangered iguanas and other conservation projects for Flora and Fauna International. Afterward, we hitchhiked back to the beginning of the trail where Jim's truck was parked. It seemed like lots of people hitched rides and everyone driving by stopped to pick people up! In the gift shop at the trail head, I bought a Saba trail T-shirt that said "You can't be lost if you don't care where you are." That slogan expressed my feelings about this trip exactly, and it remains my favorite T-shirt to this day.

The island of Saba.

On our last morning in Saba, Jim took us to the airport in the back of his truck. The passengers once more loaded up into the little plane, and the pilot taxied to the end of the runway. He then turned the plane tightly around so we were heading toward the cliff edge. He revved the engine while braking hard, then suddenly let up on the brake. We took off like a slingshot without going up — just straight off the cliff!

I wasn't scheduled to fly back to the states for another day, so we went back to the hotel on St. Eustatius. I spent some time snorkeling but the waves were pretty rough, so I mostly spent the time writing up my field notes. That evening, Jenny and I went to an Easter party on the beach but spent most of the time discussing my plans to breed the snakes in captivity. After eight days enjoying this adventure, I flew home with the snakes in my carry-on luggage, which at the time was legal since I had the official permits to carry "snakes on a plane."

This trip illustrates a trait I have observed in most of my colleagues who do fieldwork. They enjoy the challenge of working outdoors, often in less than optimal conditions. They don't ignore the difficulties that may occur. Instead, they like to solve those problems. This is particularly true in working in other countries. Risks may be involved, but being prepared to overcome those risks makes for a successful field trip.

The plane on the short Saba runway.

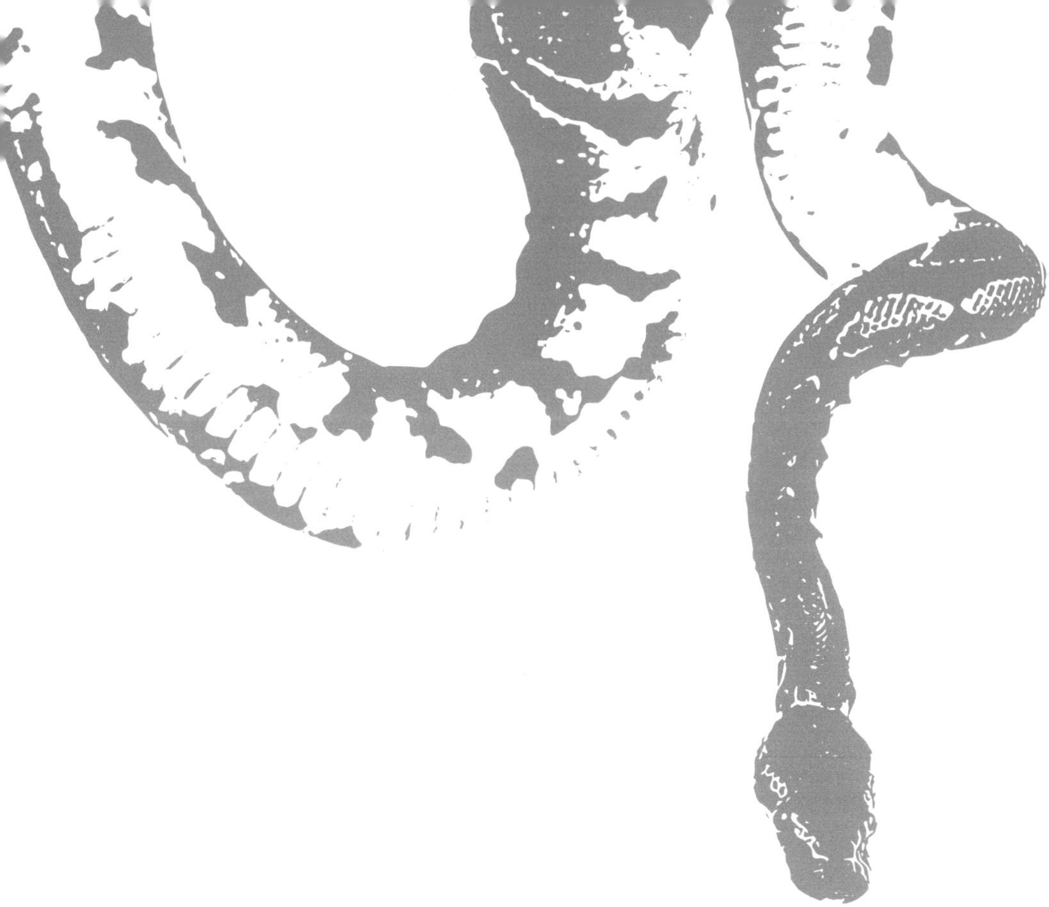

Field guides are often the starting point for new field studies as they highlight the gaps in our knowledge.

CHAPTER 10

Using Field Notes in a Book: My Trip to Study Western Species of Garter Snake

Most field guides are compilations of data collected by many researchers over many years. They include information from field notebooks and documentation on preserved specimens found in museums, in addition to data from published studies of specific species. For example, Charles Carpenter wrote a classic study of the ecology of three species of garter snakes living in the same place in Michigan (Carpenter, 1951). Field guides also use information from master's thesis and Ph.D. dissertations that have not been published elsewhere. Putting together all the known information about a group of species for a book is therefore a slow process and often takes years. However, the resulting book can be useful for many years afterward. Field guides are often the starting point for new field studies as they highlight the gaps in our knowledge.

In 1992, I suggested to my colleague Dr. Rich Seigel that we should write a book on all the species of garter snakes in North America. Originally, Rich and I were going to focus only on ecology and behavior of the members of the garter snakes, genus *Thamnophis,* but we realized the book would be incomplete without a complete taxonomy of the group. Neither of us were taxonomists, and since the phylogeny of the 30 or so species of garter snakes was very important, we recruited Dr. Douglas Rossman of Tulane to put together our current understanding of the evolution of all the species. Doug had been planning to write a garter snake book himself, so he was happy to work with us on this project. In fact, we decided he should be the first author because of the amount of taxonomic work he had done on the genus over many years. Rich and I contributed material for the behavior, ecology and physiology sections and also wrote sections on each species.

After writing a sample chapter and sending it to various publishers, we got a contract in 1993 with the University of Oklahoma Press. Since Rich and I realized that neither of us had much experience with species of garter snakes from the West Coast, we asked for an advance from the press to visit California and Oregon to see as many of those western species as we could. On that trip, we traveled to many of the most important sites where garter snakes had been studied, beginning with Henri Fitch's seminal work in the 1950s (see Henri's autobiographical sketch in Copeia, 2000). Rich had been Henri's last graduate student, so going to all of his research sites was especially exciting for him.

Rich and I flew into Reno in late May 1993, rented a great car (Toyota Camry station wagon) and headed off across the Sierra Nevada mountains to reach Eagle Lake California. Eagle Lake was a long-term study site used by famed garter snake researcher Steve Arnold and his students at Oregon State University. (Rich insisted that I note here that I was sick the whole time between Reno and Eagle Lake and he had to do all the driving.) We spent two days at Eagle Lake getting a modest number of two species of western garter snakes, *T. elegans,* and a western form of the common garter snake, *T. sirtalis.* And then we were off to the Klamath Basin National Wildlife Refuge on the California/Oregon border. We hit the motherload there for what we thought were *T. sirtalis*. We found many had swollen stomachs, so we palped out what they had eaten. We were astounded to find they had eaten voles, since most garter snake species rarely take mammals. We found more *T. sirtalis* at our next site, and from carefully reading of our western

California coastal highway.

field guide to reptiles, we realized we had collected *T. elegans,* not *T. sirtalis,* at the previous location. That species is one of the few that was known to eat mammals. Some garter snake experts we were!

Western form of T. sirtalis.

We then spent a very profitable two days with Hart Welsh and Amy Lind from the U.S. Forest Service lab in Arcata, California, right on the Pacific coast. Hart made us feel much better about our ID issues when he showed us the laminated "cheat sheet" he used to tell the multiple species of California garter snakes apart, even though he had worked there for a decade!
The best moment of our visit with them was watching aquatic garter snakes *(T. atratus)* eat coastal giant salamanders *(Dicamptodon tenebrosus)* in the

fast-moving, freezing cold streams north of Arcata. Watching Rich fall in and being swept 20 feet downstream was just a bonus!

From Arcata, we drove north on Highway 101, stopping at roadside parks and catching northwestern garter snakes *(T. ordinoides),* which seemed to be in every little patch of sunlight along the paths. Rich thought these were the prettiest of all the snakes we had seen on the trip.

Garter from Eagle Lake, California.

Our time was about up, so we headed back to Reno, the desert and, eventually, home. All in all, a very profitable and fun way to learn our western garter snakes!

Over the next two years, we gathered data on all the species of garter snakes and put together chapters on their taxonomy, evolution, ecology, behavior and physiology plus species accounts for the 30 species known at the time. Those species occurring in Canada and the U.S. had been studied at least in some locations, so our species accounts were substantial. However, the Mexican and Central American species of garter snakes were poorly known. Oftentimes, the only information available on them was from the original description of the species. I was lucky to have good contacts at the University of Kansas

The Klamath Basin on the California/Oregon border.

Museum of Natural History, where specimens of these species were deposited. The curators there were extremely helpful to me by sending any information they had on the specimens, including that from field tags and often from copies of the original collectors' field notes.

Rich photographing snakes at Klamath basin.

Members in the garter snake genus *(Thamnophis)* are likely the best known of all species of snakes because of the abundance of many of the species in the group. Our book summarized the knowledge up to that point about each of them — and the group as a whole. The book was the first in a natural history series that the University of Oklahoma press published and has been widely used as a starting point for garter snake studies ever since. For me, it was a very different experience than writing about experiments. And the book was read by a very different group of people — not just scientists interested in that specific work but by both professionals and amateurs seeking general information about the snakes.

Writing notes in a field book was a task that I learned early in my biological training. Very few students today write in a personal journal. The advent of databases for computers has put a premium on putting information in that form, not a written text. It is a shame as the skill of writing, the records of locations worked in, general information about what organisms are seen, and other information in those books are important, if only to remind an "old" biologist of the fun they had.

After you grabbed an anaconda they would wrap around you, and that was when you needed help.

CHAPTER 11 *I Felt Like a Prey Item: My Trip to the Llanos of Venezuela*

Although scientific meetings are venues for presenting of current research that is being conducted, often just as important are the impromptu conversations that occur outside of the organized talks. Those candid conversations can spark new research ideas and projects. At the annual meeting of herpetologists and ichthyologists in 1994, I heard a presentation by Jesús Rivas, a biologist with the Venezuelan office of Profauna (an official environmental organization like our U. S. Fish and Wildlife Service). Jesús was studying sexual selection in green anacondas *(Eunectes murinus)* in a nature reserve in San Fernando de Apure, Venezuela. He was using this project as the dissertation for his Ph.D. under Dr. Gordon Burghardt at the University of Tennessee. At the end of his talk, Jesús invited any interested scientists to come do research in Venezuela, but also to come give him some physical help catching the anacondas, which can get over 20 feet long and 300 pounds ("Anaconda" by J. Rivas, 2020).

I was interested in going to Venezuela because I had developed a hypothesis from earlier conversations with Bill Lamar (Tyler Zoo) and Dr. Doug Rossman (Tulane University) that a species of South American water snake *(Helicops angulatus)* might be able to switch from laying eggs to giving birth to live young. Bill had told me that he had found eggs of this species on some aquatic vegetation in Ecuador and the eggs hatched very quickly (within a week). Doug had told me about a paper he had written concerning a preserved specimen of this species that had embryos instead of eggs inside its uterus. Putting the two ideas together, I thought that this snake might be able to hold its eggs internally when conditions were too wet for eggs to hatch and instead produce live young.

I spoke with Jesús during a break in the talks and he indicated that he had seen *Helicops angulatus* in Venezuela. So we made arrangements for me to go down to his field site at Hato El Cedral to look for these water snakes during the next dry season, when he would be there working. In April of 1995, I left Tyler for a trip to the Llanos of Venezuela.

The Llanos are tropical lowland savannahs consisting of grasslands with only a few trees. They flood during the rainy season, which brings in nutrients from the rivers. This nutrient subsidy forms a huge energy base so the habitat is extremely rich and diverse in both plants and animals. I was excited to get to go see this habitat that I taught my students about in my ecology class.

I flew to Caracas where my box of minnow traps caught the Venezuelan customs inspector's eye. I had, however, learned my lesson on my trip to Haiti, so I had kept the traps in the original cardboard box and was able to let customs cut a hole in the side to check them. So this trip, my luggage only caused a little delay. However, after leaving the airport, I managed to get in the wrong taxi line, and my driver was not going to take me to the right hotel until I insisted. At the Hotel Macuto, with my "un poquito Spanish," I found out what room Bill Holstrom, the curator of herpetology from the Bronx Zoo, was in. Bill, his son Lars and Dr. Xiao-Qing Qui (whom we called Chu), a postdoctoral fellow at Rockefeller University, were all coming along at the same time.

The next day, we all flew to San Fernando in the state of Apure on the southwestern border of Venezuela. Jesús and his assistant from Profauna, Maria Munuz, picked us up. They also brought Dr. David Holzman from

Oberlin College. Dave was flying back after two weeks helping them with the anaconda project. I had previously worked with Dave on some snake learning experiments. He was low on cash and borrowed $40 from me. So then I was low on money, and we tried to cash some traveler's checks in San Fernando. But that was very difficult. After being refused by several banks, I finally found one that would change them, but I had to make three copies of each check and my passport. I also had to sign them three times each. It took over two hours, but I had learned to have lots of patience on other trips to the tropics.

We headed next to Pedro Azuja's crocodile farm to wait while Maria took the jeep, which Profauna had supplied, to get the brakes fixed. While we were at the farm, Bill helped the workers move some large Orinoco crocodiles *(Crocodylus intermedius)* out of a pond. Bill showed them how to put in breathing tubes when they taped the crocodiles' mouths shut. These animals are an endangered species, and the farm bred them in captivity for restocking purposes. They also bred endangered side-neck turtles *(Podocnemis expansis).* Jesús had arranged to keep pregnant anacondas at this farm so that when the babies were born, he could bleed them for paternity testing. While we waited, I wandered around the ponds looking at birds and saw a very large crocodile in a pool from which Pedro had said they had already been removed. Amazing how something that big can hide so well. Just like that alligator we radio-tracked back in South Carolina that was able to hide from us in that little creek (Chapter 7). Seeing this 14-foot crocodile in this pond was scary because the families of the workers had just been fishing there, and it could easily have grabbed one of the children off the bank.

We overnighted at the Gran Plaza hotel in San Fernando, but the brakes were still not fixed in the jeep the next day. So we shopped for food for the trip. We also bought cincerias (hammocks) as there were not enough beds for all of us in the ranch house where we would be staying. We finally started out for Hata Cedral Ranch but found the brakes on the jeep were still heating up. We made it to Mantical, the nearest town to the ranch, by driving slowly. We had also been driving on empty for over 20 km, so our slow speed may have actually been a good thing as we probably got better mileage. We bought gas there and ate in the only restaurant in this little town. We made it to camp before dark, but the place was a real mess and Maria was really angry at the guys from Cornell University who had just left. She got us to clean the place up while everyone alternated showering off the road dust. The shower was cold water only, but it was great to have in the house. That night, Chu and I took the one room equipped with an air conditioner. The National Geographic folks had brought it and left it behind after their visit.

The nearby city of Mantical.

The next morning, we started our hunt for anacondas. Jesús was particularly interested in getting pregnant females, so we put a 50-gallon drum to hold large snakes in the back seat of the jeep. This meant I had to ride on the roof. This was the dry season, so the flood water had returned to the streams and rivers from the flat plains. Capybaras, which are large, 200-pound rodents, traveled in herds and grazed on the grasses like our bison. The drying streams trapped fish and aquatic invertebrates in shallow pools as the water receded; and water birds were everywhere, feasting on these abundant prey. The heavy-bodied anacondas also retreated to the water, where they could move easier than on land. This also made them more concentrated for us to find. During the wet season, Jesús had to put radios on the anacondas to have any chance of finding them, since they could be anywhere.

Capybaras are large rodents that travel in herds.

Maria and I put out some of my minnow traps. She hoped they would catch baby anacondas as they rarely found the 12-inch-long babies, but I hoped the traps would catch the south American water snake, *Helicops*. Then we started walking up shallow streams (Jesús called these cañas).

We couldn't see the anacondas in the water because the hyacinth plants covered the surface, but we could feel them with our feet. The snakes were squishy when you stepped on them, and when you felt one, you just reached down and pulled it up. They only moved slowly and, despite their size, weren't aggressive until you pulled them onto the bank. Then it was important to grab them around the neck, as they would bite and although not venomous they had long sharp teeth. After you grabbed an anaconda they would wrap around you, and that was when you needed help. On the other hand, if you stepped on something that had hard skin, it would be a black caiman (related to alligators). They would violently thrash off. I stepped on one that was about 6 feet long, and it threw me completely off my feet and backward into the water.

The first anaconda I stepped on was a 2-meter-long female, and it was definitely not happy when I grabbed its neck. I was though. Jesús was also happy as it was an unmarked female. Jesús marked all the anacondas he found by clipping off a portion of the scales on the tail, in a different pattern for each individual. He said this particular female was not fat enough to be attracting males yet, but it was important to know what percentage bred each year. So it was valuable data.

Bill Holstrom placing minnow traps.

We saw lots of baby caiman swimming among the floating plants. So Maria and I caught three and put them in the minnow traps, thinking they might attract the snakes. Maria said anacondas do eat them. Although the baby anacondas apparently swam around, the adults were strictly ambush predators and would wait in the water with only their head poking out. Whenever anything happened to walk by that was the right size, they would strike and rapidly coil around it. Anacondas could eat something that weighed as much as they did. Even the 200- pound capybaras were very nervous when they were in the water, because of the anacondas. When we scared them, these rodents would dash out of the water, splashing and startling us. Walking in the same shallow water, I felt like a prey item too. I

Neil with an anaconda.

Close-up of an anaconda.

wasn't comfortable getting too far away from the group! Maria told me an anaconda had grabbed her leg the year before, and she showed me the rip in her jeans to prove it.

We generally hunted in the cooler mornings and rested in the afternoon, waiting until it started to cool down later. Or we headed over to the Aricife, a tributary of the Apure river. We would fish off the bridge there to try to catch some large rainbow bass. We mostly caught piranha. To catch piranha, we used a wire lead to prevent the fish biting through the line. At dark, the mosquitos would chase us back to the house, where Jesús grilled the fish.

There were too many bones in the piranha for me, and the spaghetti Chu fixed was very spicy.

The next morning, Gabriela, a guide from the resort on the farm, came with us. She helped me identify some of the birds: buff-necked ibis, yellow-fronted wattles, yellow-headed caracara and white-headed marsh tyrant, to name a few. Maria and I also went back to the cañas, where we had put out the traps. Nothing was in the traps, but I caught a baby anaconda on the bank, which pleased Maria. We also got two more females, so we went back to the house to process them. The ranch house had a large living area with a concrete floor, where we measured, marked and weighed the snakes. We had to run a flexible measuring tape down the back of the snakes, as the snakes were incredibly strong and "no way" could we straighten them out. We had a cattle scale hanging from the ceiling to weigh the big snakes in the drum. I was relieved to get off the jeep roof for a while, as my back was sore from bouncing up and down.

The next morning, Maria tied a foam pad on the roof of the jeep for my aching back. It didn't help much. And this time, Chu was also riding on top and kept slipping forward, pushing the pad over the windshield. I finally took it off and threw it on the back seat of the jeep. This time, we walked in some water-filled borrow ditches bordering the roads. These were created when dirt was "borrowed" to make the dirt roads higher. These ditches had more solid bottoms for walking but, unfortunately, also had predaceous diving beetles that swam up your pants and bit your legs. Bill's legs were badly chewed up, so he started using a long pole and poking for the snakes from the shore. He found a large (10-foot) male that way. The male anacondas are smaller than females and were usually less than 9 feet long.

During this outing, we spotted a small (3-foot) anaconda up on the surface of the hyacinths, which was quite unusual. Then we realized it was constricting a jacana. Jacanas are water birds with extremely long toes that allow them to walk on top of the lily pads without sinking. Birds seemed to be an important prey for the smaller anacondas, as later we found a black-crown night heron also flopping around in the hyacinths with a small anaconda wrapped around its body. This snake was an optimist, however, as it was only about 3 feet long and the heron could have just as easily eaten it!

Gabriella came over that day and told us that some tourists had seen an anaconda catch an adult deer. We checked it out and saw that the anaconda was only about 7 feet long, so it would take several hours to eat the deer. We watched for a while and then drove into Mantical to get gas and pick up John Thorbjarnarson, who was coordinating the project and had just come

Black-crowned night heron and the anaconda that tries to eat it.

Jesús, Bill and Chu catching Lydia, a big female anaconda.

in on a bus from San Fernando. Herpetology was a small world as I had met John in Haiti years prior to that. We ate at the one restaurant again, shopped for more Polar beer and headed back to see about the anaconda eating the deer. The sun was setting quickly, so we didn't stop at the house. When we returned to the site, the snake was swallowing the legs of the deer. It took six hours from start to finish. Even seven-course dinners don't take that long.

Gabriella invited us to the resort after dinner, and we met the folks from the Massachusetts Audubon Society who had seen the anaconda catch the deer. I arranged with one of them to call Joan when he got back to the states and tell her that "I'm alive." This was pre-cell phone and I had not been able to call from anywhere. The next day, the ambassador from the Netherlands and his entourage came over for a look at the anacondas, so we took a medium-size one out for them to see. Snakes always make good public relations opportunities. I always explain the important ecological role of these top predators.

Afterward, Maria and I went to check the minnow traps in one jeep, while the others went to check on the anaconda that had eaten the deer. The snake was just floating in the water, as it was so bloated. Jesús said anacondas often got sunburned when they couldn't move around after a big meal. We saw that snake floating there for three more days. We headed back to the nearby Caracaro Ranch to release some of the snakes, and while we were doing that, John T. found a group of males by poking his walking stick in the mud. Then I found the female the males were looking for. She was buried in the mud, and as I moved the mud off of her, I realized she was huge — over 13 feet long! I couldn't catch this one alone. Everyone gathered around as I pushed the mud off her head and grabbed her neck. Then it got wild as she wrapped coils around me. The rest of them unraveled her from my body. We were too far from the jeep to get the 50-gallon drum, so Jesús tied a sock over her head and she calmed down. Then we were able to carry her to the vehicle.

On the drive back, Maria saw a cow stuck in the mud that she insisted we pull out. We managed it but got so muddy that we stopped at a cattle tank and cleaned up a little. The next morning, we processed those snakes, and Jesús allowed me to name the female since she was unmarked. She was now called "Joan" after my wife. Jesús told me later that he found her every year for the next six years and she was pregnant every other year.

The next morning Maria headed into town to buy more food, and the rest of us went back to the borrow ditches along the roads. After catching a few medium-sized animals, we heard Chu and Bill shouting for us all to come over to their side. They had found another mating ball and needed help, or some males would get away. In water about a foot deep, in among the hyacinths, were five males and a very, very large female (over 18 feet and 350 pounds). We pulled the males off slowly and bagged them up. Then Chu gave his video camera to Lars to film while he caught it. Jesús remembered this huge snake because he could see the radio he had implanted under her skin two years before in 1992. He had named her Lydia after his girlfriend at the time. He said all the big females were named after old girlfriends so he wouldn't run out of names. It took all five of us to carry this girl to the jeep.

That night it started to rain, which was early for the rainy season. So the next day, I sat on top of the jeep with my umbrella open. Everyone laughed and called me Neilly Poppins, but I stayed dry. We didn't see any snakes that morning, probably because of the clouds, so we took the rest of the day off. Bill played his little trumpet and Jesús sang some Spanish songs. He had a great voice. I showed them some Tai Chi, which Chu said he learned as part of his medical training during the cultural revolution.

When the rain stopped, we drove out to the Caracara River (Nodulo los Rayados) to fish. We caught some Jewel Cichlids, which were a beautiful gold and blue. Joan kept cichlids in aquaria back home and would not have believed we were going to eat these! These fish were larger than the piranha, with fewer bones, making the meal much more pleasant. Maria served

Muscatel with dinner, which probably helped. After dinner, we talked about her research project with the baby anacondas. Her English was about as good as my "un poquito" Spanish, but we got things across. Discussing projects with other biologists was always an important aspect of traveling to field sites, probably because there was always some downtime.

The next morning, one of the jeeps had a flat. It reminded me of all the flats we had in Haiti, but we couldn't blame the street urchins here. We only had one spare tire between the two jeeps and about half the parts of a jack, but Jesús got it changed. We also had to siphon gas into the other jeep from the drum of gas that Maria had brought back. Maria made the rest of us clean up the house while Jesús filled the jeeps. She implied all gringos were slobs, but I didn't see Jesús doing any cleaning.

One day, we drove across the savannah to the Quaratarito River, which, because of its current, did not have the hyacinths covering the surface of the water. Jesús said this was where he thought the pregnant females came, because they could warm their embryos while hiding from predators among the roots of the short mangrove trees along the banks. I can't imagine what would prey on an adult anaconda. But maybe Orinoco crocodiles could? We checked out caves and holes in the banks and found an anaconda about 14 feet long, hiding back under the roots of a tree. We poked it several times and made it angry. So it darted out into the river. Bill grabbed its tail as it entered the water, and he held on while Jesús tried to find its head as it was splashing wildly. When I show Chu's video of this capture to my students, they always gasp when the snake lunges at Jesús and almost bites him. I turn the sound off as they don't need to hear our bad language.

The mornings of the second week were pretty much like the first week's efforts. Then, each afternoon, we processed any snakes we caught and took them back to be released. We did hunt one day for an animal in which Jesús had put a functional radio. Even with the radio, it took us a long time to find her in the pool we knew she was in. You wouldn't think it would be hard to find an 18-foot snake that had a radio! Much like that Orinoco crocodile at the crocodile farm, she was nearly invisible in the murky water. Since we found seven males near her, she had likely mated. So we took her to an outdoor enclosure where Jesús could keep her until she had her babies. This tank had been used to head-start baby crocodiles. When we put her in, she crawled straight to the cement pool in the center of the cage. She was very heavy, so she would probably have a large litter; maybe 45 babies. Jesús drew

Processing snakes.

blood from all the males around her and would draw blood from each baby to evaluate whether the female had mated with more than one of the males. Knowing whether the females multiple-mated was important in determining whether a commercial harvest could be sustained on these animals.

The tourist camp officials loaned us a sight-seeing boat, so after lunch one day, we headed up the Matiyure River. We saw lots of black caiman on the banks but no Orinoco crocodiles. We saw huge numbers of birds, including Orinoco geese, Amazon kingfishers, sunbitterns, rufescent tiger herons and hoatzins. This refuge is a major birding attraction for tourists. Seeing hoatzins was particularly interesting to me because, in my classes, I explained that they were one of the few bird species that ate leaves. Like cows, hoatzins have a ruminant-like stomach (four chambers) with bacteria that help digest the cellulose of the plants. The young hoatzins retained claws on their thumbs to help them climb the short mangrove trees, since the leaves made them heavy. At the end of this river, the water spread out into the savannah, and in the setting sun, we saw millions of shorebirds. One great thing about going to other countries to hunt snakes is you also get to see incredible sights that others pay lots of money to see.

Maria and I went up to one of the guard houses and bought some chickens for the caged anaconda. This house had a vegetable garden, which Maria said was unusual. Most families out in the country didn't do gardens, since they could pick wild fruits almost everywhere. This house had a room where they sold the vegetables. In the corner, it had a large clay jar, which Maria explained was the traditional Venezuelan way to keep water fresh. The outside of the jar was covered in moss and ferns!

We took the jeep for one last look at the anaconda that had eaten the deer. This was our last day to hunt, so any snakes we caught we had to process that day. Jesús wanted to mark this snake but it was still floating in the same spot, so we decided not to bother it. Bothering it might have caused it to throw up the deer. Jesús had seen snakes die from injuries to the

Happy catching anacondas.

mouth when they did this. Maria and I made a final walk through a hyacinth-filled pool, and on our last pass, a snake brushed her leg and she grabbed its tail. I also grabbed the snake, and it pulled us both down into the water. So, I'm sitting on my butt in the water with a snake by the tail and I couldn't see the "business" end of it. We yelled for help and the others finally arrived (probably only minutes later). They pulled the snake out on the bank, and I grabbed it's neck. Yeah! This was number 40 for the three weeks. Of course, we didn't have a barrel with us, so Lars donated a sock to put over the snake's head.

I packed up that evening but left most of my clothes for whomever came next. They were so dirty, they would not come clean anyway. I also left the minnow traps for Maria to use. I slept in a cinceron that last night, and even wrapped in sheets, I got cold. The others flew out from the San Fernando airport the next morning, but I stayed overnight in Caracas, at a hotel Jesús recommended, La Quince Letras in Macuto. It was very nice with a restaurant and a walkway to the ocean. I called Joan that night, after not being able to for 16 days. Although I didn't catch any *Helicops* on this trip, I made some important contacts and friends, learned a lot about the Llanos and caught lots of anacondas! Later, Chu sent me four hours of video that I cut down to 20 exciting minutes and showed to my classes for several decades afterward.

I had heard that snakes were abundant in Trinidad since it was an island with fewer predators and near enough to South America to have a diverse snake fauna.

CHAPTER 12

Why You Need Two Trips to the Same Place: Collecting Water Snakes in Trinidad

I had not been able to find the South American water snake *(Helicops)* in Venezuela. I wanted to study this species to see if it might be able to switch from laying eggs to having live young. I had heard that snakes were abundant in Trinidad since it was an island with fewer predators and near enough to South America to have a diverse snake fauna. Islands are usually fairly depauperate, as I found on St. Eustasius. I took two trips to Trinidad as *Helicops angulatus* occurred on that island. My first trip was to scout out field sites and also to obtain the collecting permits for the snakes. I had not been able to secure the permits ahead of time, so I planned to go to their wildlife agency in person.

I took the first trip to Trinidad in March of 2000. I had been trying to get permits by email and phone calls for several months without luck, so I planned to just go to the environmental office myself. I decided to make it somewhat of a vacation, so Joan and my son, David, came along. We stayed over in Port-a-Spain, at the Bel Air hotel near the airport, as we had booked a car for our next morning. Of course, it turned out this was just after Carnival, so no one got up early and the car wasn't brought to the hotel until the afternoon. So we looked at birds around the hotel, mostly those common in the city like palm tanagers, great kiskadee, carib mockingbird and white-winged swallow.

Trinidad had a historical tie to England, so people drove on the "wrong" side of the road and, as in most tropical countries, also very close together. Joan got up the courage to drive first, so I navigated. Roads that looked huge on a map were tiny in reality. After several wrong turns, we finally made it to the town of Blanchisseuse. We had trouble finding our hotel, as there were few street signs. We finally found it and checked in at Hotel Bel Mar. Afterward, we went down and swam in the bay. We had to stay shallow, though, as there was a strong rip current. We almost lost David in one. We went back for dinner as the sunlight faded, but the electricity was out in the hotel. They were used to this, and the cook managed to fix fish soup for us and shrimp and rice for David.

The next day, Joan and I went birding down the road, which went over a picturesque old wooden and iron rail bridge. After breakfast, we explored more. Joan went birding again, and David and I went to catch the *Ameiva ameiva* lizards that were on the forest edge. They were fast and most got away into the forest. Later, we caught a gecko, *Gonatodes vittatus,* on a concrete wall near the hotel. David and I went out that night, down the road, but it was very dark and we didn't feel comfortable going too far into the forest. So we had no luck there. We did get lots of geckos on the hotel walls. Many gecko species have adapted to live mutualistically with humans.

A gecko.

When we checked out the next morning, the whole bill was just 448 TTs ($75). As we left in the car, the brake light came on, so we had to get the rental company to come out and fix it. Which they did quickly. We headed down the coast road, very narrow with a steep rock wall and forest on one side and ocean cliff on the other. We did make it back to Port-a-Spain and went to the Zoo to meet the curator, Sayeed Ali, who was going to take us to the wildlife office. While we waited for him, a keeper showed us a huge amphisbaena (called worm lizards although they are not lizards). On the drive to the wildlife office, we stopped at a large intersection in the highway that had street vendors, and Sayeed bought some traditional Trinidad "doubles" for us. These small burrito-like sandwiches were made with chickpeas, mango and peppers on a fried cornmeal wrap. He bought a mild one for Joan but got it mixed up with my spicy one. It was painfully hot for her. Sayeed bought her a coconut, because coconut milk would help cut the burn.

Zookeeper at the Emperor Valley Zoo in Trinidad with an *Amphisbaena alba*.

At the wildlife office, they did not have any of my letters or notes of my phone calls to them. So they just had me write a letter as to why I wanted the snakes. But after I did that, they said the snakes were a protected species and I couldn't get a permit. Dejected, we were getting ready to leave as David Chadee, a wildlife officer, drove up with folks from the Cincinnati Zoo who were working on a reintroduction project with blue-and-yellow macaws. The zoo folks knew who I was, and that changed the tenor of the meeting a little.

Mr. Chadee also said the species were protected and I probably couldn't get a permit. However, he hedged some and invited us to go with them the next day to release the birds. I arranged to meet them the next day, thinking that might be a way to get his help with the permits.

We drove up to Pax Guest House at Mt. Saint Benedicts, where we had arranged to stay. We had to ring a bell for the monks who ran the hotel to let us in. The mission of this order was to help travelers, so this hotel was part of that. Our rooms had a balcony overlooking bird feeders of several types, and there were all sorts of birds there. It appeared that all the other visitors were bird watchers, including an ichthyologist from Oxford on the next balcony over from us.

We hiked up to the forest to a donkey trail that was supposed to be good for birds, but it was hard to see in the thick forest habitat. I did catch a poison dart frog *(Mannophyrne)*. We had much better luck with birding just off the terrace, above the feeders. We saw lots of hummingbirds. The white-chested greens and black-throated mangos buzzed all around us. Jarod, the hotel owner, told us there were frogs living in the pond out front, so David and I went there after dark and caught one and also a gecko, and photographed both. Dinner that night was pea soup, fried kingfish and rice, plantains, and strawberry surprise for dessert.

David on donkey trail.

The next day, we drove to the wildlife office and, low and behold, they had a permit for 12 *Helicops* for me! David Chadee said we wouldn't be going to let the birds out until later, so Joan, David and I headed over to Caroni Swamp for a boat ride to see a scarlet ibis rookery.

This swamp was a very large mangrove estuary. We checked out some ponds in the area first, where we saw white-headed marsh tyrants and other marsh birds. We then went over for the tour. Our boat had five German tourists. As we rode along the canal, I spotted a tree boa *(Corallus ruschenbergerii),* and the boat driver got close enough for me to grab it. He was surprised, as were the tourists, who all scrambled away from me. I explained that it was harmless and they moved back, at least a little. I let the snake go back into the mangrove trees. We reached the mud flats where the swamp met the ocean. As thousands of birds flew in at dusk, the trees looked like Christmas trees with red ornaments.

Neil with tree boa.

The next day, after a few wrong turns, including going the wrong way on a one-way road, we found the furniture store in Sangre Grande. We met David Chadee there to go to the watermelon farm where the birds were being kept. He drove our car, and the zoo biologists brought their Range Rover. It was a wild ride along the eastern coast, where much of the road was damaged from recent rains. We turned into a large estate along the beach that had a large aviary containing the blue-and-yellow macaws. These birds had been collected in Guyana and were to be released in Nariva Swamp, which had a large nature preserve called Bush Bush. One of the farm workers said they just released a water mabuda *(Helicops)* they found in a watermelon farm just down the

South American Water Snake (*Helicops angulatus*).

road. I went over with them and found it in an irrigation ditch. I caught my first *Helicops!*

We all drove down to a house next to a small channel to wait until the tide came in before entering the wildlife sanctuary. I bought gas for the little flat-bottom boat and we loaded up with the two Cincinnati Zoo guys, David Chadee's assistant, Sham (everyone called him "Short Man"), and David, Joan and myself. They put the loudly squawking birds in a cardboard beer box on David's lap! We then plugged slowly down the channel because of the small clearance the boat had above the waterline. We were dropped off on a grassy hillock and walked down a trail to the forest. There was a little shack here that even had a bathroom! Steve, one of the zoo biologists, said he had stayed there overnight and all 95 species of mosquito in Trinidad bit him!

Further down the trail was a large aviary for the birds. They would be given a few days to acclimate before being released. This was called a soft release and increased the chances of the birds surviving. We hiked down the trail and saw a white-faced capuchin monkey with two babies in the canopy. Further down was a tree higher than the forest with a watchtower at the top. David and I climbed it. It was about 110 feet high with a metal ladder attached to the side. It had outer rings partway up. The uppermost part of the ladder did not! At the top, we had to use our elbows to lift ourselves through a trapdoor and into the observation treehouse. The view over the top of the forest was worth the effort and the risk.

Short Man then led us out of the preserve, keeping a pretty fast pace. We stopped for some red howler monkeys moving through the trees above us.

We also saw channel-billed toucans. We ran out of water, but just as we started to worry, we entered a small clearing and the jeep was there! Short Man had known exactly where he was going! We drove back from there, picked up our car and returned to Pax Guest House. The rum punch as soon as we got in was much appreciated! I took the snake over to show to a grade school class that was visiting.

David Chadee with blue and yellow macaw.

Next day, we headed to Asa Wright Nature Center. I had known about this place for years, from the books of William Beebe and because I had read that the largest boa constrictor ever found came from here. There were lots of big mining trucks on the road en route to the center. We missed our turn to Arima bypass twice, so we had to ask for help to Blanchisseuse. We finally made it. We paid our $6 each and went out on the veranda of the center. Just below us, they had put out fruits, and a golden tegu (large lizard) and an agouti (large forest rodent) were feeding on them.

At the bird feeders, we saw honeycreepers, bearded bellbirds and a different hummingbird. We went on a tour of the grounds and saw manakins at a lek (male display area). We ate lunch on the veranda and birded a little more before heading to Mt. Plasair Estate and Spa, where we hoped to see sea turtles. At the town of Valencia, we headed north and the road was very narrow. The coast was beautiful, but the towns all looked pretty rundown. We wondered what the hotel would be like. We arrived at the town of Mt. Grand Riviere, which was quite small. We took a little dirt road and pulled into a backyard. Yes, this was it! We checked on the cabin and it was not bad. It had a fan and there was mosquito netting over the beds. They had a dining area for supper, which was pumpkin soup and grilled shrimp. We arranged to be woken up if any turtles were spotted during the night. The cabin had a large, open window, and as we got ready for bed, a skinny white horse stuck his head in.

About 2 a.m., the guide, Francisco, woke us up through the open window. We hiked about a quarter mile down the beach to join a group that was watching a turtle dig a nest. Francisco was telling the group about leatherback sea turtles. This one was huge! It must have been 12 feet long. We weren't allowed to take pictures as it might have disturbed her. She laid about 75 eggs. Then she dug some fake nests and headed back to the ocean.

While walking the beach the next day, we saw several turtle nests that had been dug up by dogs. We didn't understand why they didn't protect the nests! We drove to a lighthouse on the point of Toco. The drive along the eastern coast, through forests and rivers, was very pretty. In the town of Sangre Grande, we found a phone and called David Chadee to see if Short Man had found any water snakes for me. But he said Short Man's watermelon patch had caught fire so he had not been able to look for them. Back at Pax Guest House, a group was playing steel-drum music for some kids from New York who were staying there after snorkeling on Tobago. David Chadee called me, and we talked about how I might get the snakes on a future trip. We flew home the next day.

Second Trip to Trinidad: Snakes on a Plane Again

I obtained funding from the conservation fund of the Cameron Park Zoo in Waco to go back to Trinidad to collect the water snake, *Helicops angulatus*. Johnny Binder and Terri Cox from the zoo came along as did my son, David. David and I missed our plane in Miami after our flight out of Dallas was delayed. I found a different flight to Trinidad, but it was late when we arrived and Johnny and Terri were gone. We took a taxi to Pax Guest House and had to wake Gerard, the manager, to let us in.

We met Johnny and Terri at breakfast the next morning, and they wanted to go on the Caroni Swamp birding tour that evening. So David and I went up to the Old Donkey Trail for a while. That afternoon, we took Johnny's rental truck into Tunipuno to get supplies, but it was Boxing Day so nothing was open. At the boat dock for the tour, we watched some large four-eyed mud skippers swimming around. On the boat ride out to the scarlet ibis roost, we saw several Cook's tree boas in the mangrove trees, and one was low enough for me to catch. It was about 4 feet long and bit my arm. The other people in the boat were a bit surprised, to say the least. Terri videotaped the whole thing. The rookery was amazing, mostly scarlet ibis but also tricolor herons and cormorants.

Back at the guest house, before we headed to bed, David and I caught a gecko, *Thecodactylus rapicauda,* a frog, *Mannophryne trinitatis,* and saw an elf owl. We called Short Man the next morning and arranged to meet at Courts furniture store in Sangre Grande. It was the Muslim holiday, Eid Day, so few stores were open, but we found one and got some lunch. Short Man put our gear in the bed of the truck, and David Boodoo was inside with David and me. I think David was the same guy I was calling David Chadee last trip. He had my permits, but I had to write out my request again, even though I had faxed it twice. He took us to a three-bedroom house we rented. It had a kitchen and shower! We thought we were getting a one-room hut. I did have to buy some office furniture for the wildlife office and a veterinary certificate for the snakes. All in all, it ended up costing about $50 a night.

We got our gear together and headed down toward Nariva Swamp. Johnny loved driving the road, as half of it was washed away. Short Man told us that 6 miles of the road had been under water for a week. We took the four-wheel-drive truck up a muddy road and into an irrigated watermelon farm. There were ditches everywhere, all the way up to the mangrove forest. So that first day, we put out the minnow traps that I had brought. Terri videotaped it all for a TV program for the zoo. As the sun went down,

Our house at Sangre Grande.

the mosquitoes came out in droves! As it grew dark, David and I hunted but were eaten alive, so we headed back. Johnny and Terri had four snakes already. Then Short Man showed up with 31! We think he might have been hunting on the nights before we showed up. Back at our rental house, we cleaned up and David Boodoo invited us to dinner at his house, just across the street. We had a traditional meal with a tamale-like dish and macaroni pie.

The next morning, I fixed coffee on the one-burner propane stove. It was wonderful! I loved fresh coffee in the tropics. Short Man showed up, and we took him to join a tour group from Canada that he planned to show around Bush Bush preserve. We stopped for cigarettes for Short Man at Duggies Meat Market and then dropped him off at the house by the canal where boats were kept for traveling into the refuge. We went back to the watermelon farm and started checking traps. The fifth one we opened had a snake, and we were thrilled. The next 20 traps had eight snakes, cymbranchid eels, big tadpoles, a metamorphosed paradoxical frog and lots of cichlid fish. Paradoxical frogs are called that because the frog metamorphoses much smaller than the tadpole.

The 31st trap had a *Hydrops* (false water coral snake) which was a new species for us. We got a total of 15 snakes in 33 traps! This was almost a 50% capture rate, whereas 10% was very good for minnow traps normally. As the tide started moving out, the ditches with traps near the mangroves were getting dry. So we moved those traps closer to where we had parked. Then we walked over to the beach across the road for lunch.

Terry Cox and Short Man.

After lunch, we headed down to the canal house, but Short Man was not back yet. So we drove to a small town nearby and got some snacks and water. Heading back, Johnny yelled that something else had been eating his crackers! They had roaches! I told him it was just extra protein but gave him one of mine to get the taste of them out of his mouth. At the canal house, the first load of tourists had come in and my son, David, showed them the snakes. When Short Man showed up, we headed into Sangre Grande to get Roti for dinner. This breaded sandwich was full of a curry mix of chicken bones and not so great!

A paradoxical frog.

Johnny Binder checking traps.

But we could at least say we had eaten Roti. We went to the house and I sexed and measured the snakes. The others walked to town to buy a fan as Terri's hadn't worked the previous night and the mosquitoes had been bad. Short Man came by, but we were too tired to go night hunting. He told us of a good Chinese restaurant for the following night. After dinner, I called Joan from David Boodoo's house.

We got up early and Short Man had brought doubles for us for breakfast! We headed back to the watermelon farm to check traps, and we ate the doubles at the grass hut there. Holy cow, they were spicy! I almost died. In the traps, we got 14 more *Helicops* and one *Erythrolamprus.* The bigger females were in the ditches closest to the forest, so we moved some traps. We headed down to the canal house and headed into Bush Bush. The water was higher than the last time I had gone, so we were able to take the boat all the way to the little cabin in the forest. Short Man knocked some coconuts out of a tree, and we drank the milk and then started hiking the trail. We saw red howler monkeys in the canopy. Short Man took us to the largest tree in the forest, which he said he did not normally show tourists. David and I saw a big black and yellow snake *(Pseustes)* and a prehensile-tailed porcupine (yellow-bellied). Short Man asked if we were ready to head back. We were exhausted and all shouted, "Yes!" He said suggested we go just a little farther, but I remembered he tricked us the last time. I knew we were nearly back to the beginning of the trail. Ha ha! When we returned to Sangre Grande, Terri and Johnny went to a bank and then picked up pizzas at Marrios. We were too tired to go night hunting again. Instead, we went to Short Man's house and had Bundt cake, cookies and chocolate that his wife had made. David Boodoo came by with my permits! Yay. He had written them for more than I asked. Because the snakes were so abundant, I guess. I started picking out snakes to take.

The next morning, David Boodoo and Short Man came over and we headed to town to get the Dutch couple David was taking to Bush Bush preserve. David bought us doubles for breakfast (four each for 20 cents). His wife packed us lunch. At the watermelon farm, we had a bet of a six-pack of beer for how many snakes the traps would have. We got 25 *Helicops* and one *Hydrops,* the water coral, in 40 traps. Short Man won. We also caught more cichlid fish, which the trapped snakes had gorged on. And we got some Surinam toads, *Pipa pipa.* An interesting characteristic of these aquatic frogs is that the father carries the babies, embedded in his back.

We went over to the Manzinella beach and ate our lunch, consisting of cut-up chicken parts in rice. They seemed to just hack up their chicken any old way. The Dutch couple also was having a picnic lunch there. After lunch, we headed up to a nearby Caribbean Manatee Sanctuary. We took a rowboat out into the Nariva River to look for them. We saw a huge ringed kingfisher but no manatees. There was a treehouse that provided a view of the manatees feeding along the shore. Terri was afraid of heights but climbed up anyway. We could see where the manatees had eaten grass along the shoreline. We then headed to Brigand Hill Lighthouse. The road was steep, and the four-wheel drive barely made it. A guard let us in and we climbed up the lighthouse, where we could see all of Nariva swamp. On the way back to our house in Sange Grande, we stopped at a feedstore and got Johnny and David big cutlasses and leather sheaths — because every 16-year old needed a cutlass, right? Short Man said his father would cut them down to machete size and sharpen them for us.

We were not as tired this evening, so we headed back to the watermelon farm to hunt the ditches for snakes. I wore a mosquito-net hat and long sleeves, but the water was much lower this time and not many mosquitos were out. Not many snakes either. Terri and Johnny caught a rainbow boa *(Epicrates maurus)* but no *Helicops.* Short Man only caught eight. We went to his prime spot, which had less vegetation so the snakes were easier to see in the water. We got four more but then called it a night.

We got up early to go to Bush Bush to see the red-bellied macaws take flight from their roost in Bush Bush reserve. At the canal, we found that another group was unloading kayaks. They were inexperienced and kept crashing into the banks, blocking our passage. By the time we got halfway into the preserve, we saw the macaws taking off. So we turned around. We headed over to the watermelon farm and pulled our traps, as this was our last day. We got 15 more *Helicops* for a total of 104. The snakes' bellies varied from white to red, and David and I had been noting the venter color of the snakes we caught. We later published this variation in the Journal of the Caribbean Biology (Ford and Ford, 2002). We stacked up the traps, and as we carried them out, we encountered a big yellow-bellied puffing snake, *Pseustes sulphureus,* and David caught a vine snake, *Oxybelis aeneus,* in the

Yellow-bellied puffing snake.

grass. We processed the last snakes at the house, and David Boodoo's wife had us over for lunch. Since it was New Year's Eve, all the restaurants were closed. But Short Man's family came over to see the snakes, and they invited us to come over that evening.

That evening, we walked around town to all the neighbor's houses, and everyone was cooking outside. We were fed everywhere. There also was a lot of drinking, even though most of the people were Hindu. There was a Hindu school right next to our house. David was given a Shandi ginger beer, which had 1% alcohol. There was loud music everywhere too. Short Man invited us over to his house. I headed back to our rental house, but David went over there with Johnny and Terri. David showed up at our house at 2:30 a.m. and fell in bed with his clothes on.

I was up by 7 a.m., when Short Man was expected to arrive. But he didn't make it. Surprise, surprise. David Boodoo brought over my final permit, and because the snakes were so abundant, he had changed the number from 12 to 20 animals! We went to release the extra snakes and ran across a trap we missed. It had another *Hydrops,* the water coral. Back at the house, Short Man brought David the sharpened machete. It was so sharp you could shave with it! We had KFC for lunch and packed up to leave.

We put all the snakes in our two carry-on backpacks for the flight, but about halfway to Miami, the snakes started defecating. They had eaten huge amounts of fish in the minnow traps and had digested them by now. They really stunk, so we took turns going into the plane's bathroom to wash out the snake bags. That helped some with the smell, but I imagine our close neighbors on the planes were not happy. We had to go through the agriculture line in Miami, but they just passed us through when they saw the permits. In those days, you could take "snakes on a plane" without any trouble.

“I was able to complete a project in just three days.”

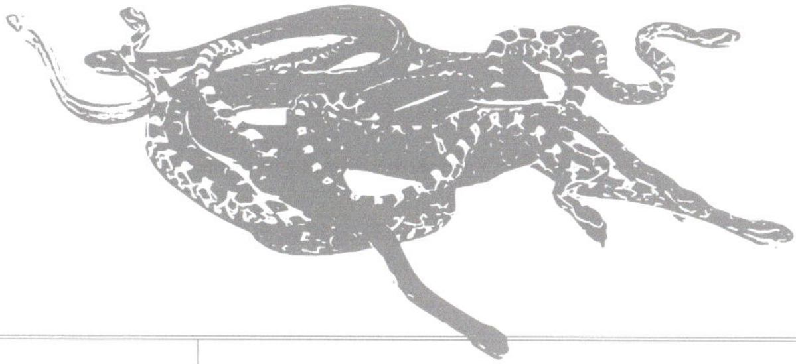

CHAPTER 13

Field Projects Don't Have to be Long: My Trip to the Snake Dens in Manitoba, Canada

I have indicated in previous chapters that field biology projects generally take a long time to complete because of the nature of the work. The collection of field data requires trips to sites that may be remote, i.e. on Caribbean coral reefs or ice flows in Antarctica. The projects are often dependent on the correct weather or seasonally restricted, i.e. mussel surveys in rivers that flood in the winter. Sometimes the organisms' own behavior can delay work, as when their activity patterns are restricted, i.e. only being active at dawn or dusk (crepuscular) or living mostly underground (fossorial). However, some places and projects are the exception to this, and projects can be accomplished quickly. This happened to me when I was invited to the garter snake dens in Manitoba, where I was able to complete a project in just three days.

For more than 30 years or so, several biologists from different universities have been studying the behavior and ecology of the red-sided garter snakes, *Thamnophis sirtalis,* that den in limestone sinkholes in Manitoba. Because summer in Manitoba is of such short duration, the snakes must mate immediately after coming out of hibernation in order for the females to have enough time to develop their babies internally. Most temperate-zone snakes in lower latitudes do not mate until several weeks or months after they emerge from hibernation. Because of the limited number of suitable dens deep enough for them to hibernate in without freezing, the Manitoba garter snakes aggregate together in huge numbers at those sites. This aggregation of snakes allows them to find mates quickly but has also produced extreme competition among males for females.

Because of my work with garter snakes over the years, I had always wanted to see this remarkable phenomenon in person. So in 2002, I asked Dr. Bob Mason from Oregon State University if I could come visit while they were there working at the dens. Bob was the principal director of the research conducted near the town of Narcisse and coordinated the work on red-sided garter snakes there. Bob not only encouraged me to come visit the dens, but he said I should develop a project that we could do together.

Bob picked me up on May 12 at the Winnipeg airport, along with Dr. Randy Krohmer, who had come from Saint Xavier University. We stayed in a house that Bob had rented in the nearby town of Chatfield. Dr. Rick Shine from the University of Sydney was already there working, along with

Bob Mason and Rick Shine.

several of Bob's graduate students and an Earthwatch volunteer. We had many discussions about projects with snakes in that house. But it seemed like every one I came up with, Rick had already done back in Australia. He was an incredibly prolific researcher!

This year, their research was being conducted at the dens near the town of Inwood rather than the big dens at Narcisse. Narcisse had become popular with visitors, and they sometimes slowed down the research. The scientists didn't mind answering questions, but since the window of snake activity was very short, it was critical for them to work as quickly as possible.

It was still very cold the days I was at Inwood, but there were plenty of snakes coming out of the den every day. Male snakes came up when even a little bit of sun broke through the clouds. The males would hang around the surface rocks warming up, and when individual females came out, whole groups of males would start courting them. Sometimes as many as 50 males would be on the back of a single female, courting her. I took lots of video of the snakes courting since I really wasn't that involved in the main field research. One interesting study that Bob and Rick conducted showed that some males produced the female chemical attractant (pheromone) to get other males to court him, thus warming up that male quicker and giving it an advantage in mating.

After some discussion the first night over beers, Ryan O'Donnell, one of Bob's students, and I decided to do a pheromone trailing project together. We built a Y maze at the house to test whether male snakes could distinguish the trails of females that had already mated versus females that were unmated. During copulation, the successful males produced a mating plug that functioned like a chastity belt and prevented the female from mating with other males. We hypothesized that males would not waste energy trailing a female that had already mated, since they would not be able to mate with her.

Each morning, Ryan and I would go to the den and collect males and females to test. We put half the females into an outside arena with males until we saw them mate. We kept the other females isolated, so they were unmated. We had females in each condition crawl down different wings of the Y maze, and then we released a male at the entrance of the maze to see which trail they would follow. We thought the males would choose the unmated female trail, and sure enough, they entered whichever wing contained that trail and avoided the trails of mated females. Using fresh males after each test, we had enough data within three days to publish our results (O'Donnell, Ford, Shine and Mason, 2004). So much for fieldwork with snakes taking a long time to conduct.

Later in the week, Bob took me to a couple of other dens, the Narrows and a giant quarry nearby, but we didn't see as many snakes at either site. They had already begun dispersing from the dens. The snakes traveled out to grassy fields miles away, where they fed on earthworms and frogs for the short time until they delivered their babies.

Although snake dens are found throughout the northern parts of North America, the Manitoba dens are very unusual because of the sheer numbers of animals, tens of thousands in single dens. Field research often takes multiple years if not decades to conduct. In these Manitoba dens, hypotheses could be tested quickly because of the ease of access to large sample sizes. This does not mean that this project overall was a short-term one. Each experiment produced a better understanding of how reproduction was controlled in these animals, and the results

The Inwood den.

generally led to more questions. For example, Randy's research dealt with how chemicals in the brain of the male snakes were involved in triggering the courtship behavior. Normally, testosterone is the hormone that controls male behavior, with lengthening light cycles causing males to begin to produce testosterone. However, Randy found the level of testosterone was low in the blood of these garter snakes when they left the den. Through a series of projects over the years, Randy teased apart how changes in the brain of the snakes during the cold temperatures in hibernation stimulated the males' behavior when they warmed up in the spring. Each experiment Randy conducted was accomplished relatively quickly, but the final answer to his question, i.e. what stimulates reproductive behavior in these animals, took several years to determine.

These researchers at the Manitoba snake dens have produced a large body of work as to how reproduction is controlled. How that information translates to snakes in areas with more normal seasons has required additional research. In particular, we still have much to learn about the reproductive physiology of snakes in tropical habitats with just wet and dry seasons. And because of the rarity of snakes there, the research is much slower.

Garter snake mating ball.

I ended up being the 'snake wrangler' and helped them get the shots they wanted.

CHAPTER 14

Talking to the Media: Discovery Channel TV Show at the Old Sabine Bottom

When you work with snakes, it is quite common for your work to get media attention as most people have strong responses to snakes in one way or another. For me, generally this interaction involved local newspapers or television, but occasionally it was national media or TV. Because I had published several scientific papers on reproduction in cottonmouth snakes, I was approached by a filmmaker for the Discovery Channel and asked if I could get females of this species that would have their young so they could film it. They had recently produced a film that included an anaconda giving birth and were now making a movie about venomous snakes. They wanted to film one giving birth.

When they contacted me, I indicated that it would be difficult to time the births such that they could plan to be there at the right time. However, they really wanted to try. I suggested I knew about the general time of year that cottonmouths would give birth and could collect and hold pregnant females for them. Their film crew then came to Tyler for a 10-day period around that window of time (late August of 2002). However, with animals, things rarely go smoothly and the pregnant cottonmouths I had collected all gave birth the day before the film crew came to Tyler. We did stage some shots with the babies and a mother at the Old Sabine Bottom

Wildlife Management Area where I had collected the pregnant females. North American vipers would stay with their babies for a few days after they were born, so although staged, this was a realistic piece for the video. The filmmakers were also able to get the birth of babies of a copperhead that I had in my lab. When I went to check on her, I saw she had delivered one or two babies, so we rushed her and the babies outside, next to a tree. The film crew came over quickly, and she delivered the next eight on cue. I think they were happy with the time they spent here as it became the first 10 minutes of their show "Tracking Deadly Vipers." For me, it was interesting to see how they staged the scenes for the program. I ended up being the "snake wrangler" and helped them get the shots they wanted by keeping the snakes under control and putting them where they wanted them to be. It took a lot of repeated tries to get just a short segment of film.

Discovery Channel crew filming at the Old Sabine Bottom Wildlife Management Area.

After the show aired, it was interesting to see how others reacted to my part in this movie, relative to how my scientific papers were received. My scientific papers were typically read only by other scientists, but even the administrators of my university thought this show was noteworthy and frequently mentioned that I had been on the Discovery Channel. Although I had always found time to talk to nonscientists, this experience reinforced to me that educating the public about snakes and the value of understanding them was an important function of a field biologist. This activity was not one for which we received training, but with snakes it was easy to hold an audience.

My work with cottonmouths started by "accident." With a reputation for aggressive behavior, they were not the most attractive or appealing snake and I had not planned any projects with them. However, they turned out to be very easy subjects to study in the field. Cottonmouths were common in aquatic habitats in East Texas, and they were easy to catch since they did not tend to try to get away from humans. Surprisingly, they weren't particularly aggressive either. Pregnant females were easy to find, whereas in most other species of snake, the females were inclined to hide while pregnant. I would walk around drying pools in lowland areas and the females would be in ambush poses along the edge, waiting for prey to come by. I also happened to have access to two very different field sites where cottonmouths were very abundant, the Old Sabine Bottom Wildlife Management Area run by Texas Parks and Wildlife and Sheff's Wood, a preserve owned by the Texas Nature Conservancy.

Cottonmouth at the Old Sabine Bottom.

Sheff's Wood, where I studied snakes for 14 years, was a small Nature Conservancy preserve near Tyler State Park. The Old Sabine Bottom Wildlife Management area, where I worked for the next 15 years, was a large Texas Parks and Wildlife property just north of Lindale, Texas, on the Sabine River. I took over the management of Sheff's Wood in 1984. It had been established in 1968 as the first Nature Conservancy preserve in Texas. It was a 74-acre mixed hardwood and pine forest with a small creek through a bottomland. I got a small grant from the Conservancy to put in drift fences and do baseline surveys of the snakes at Sheff's Wood. These were 50-foot-long sheet metal fences that led to funnel traps. The snakes would follow the fence and get trapped. It took some serious digging to get the fences in the ground. Vince Cobb (now a professor at Middle Tennessee State University), who was my grad student at the time, and a number of undergrad students helped me dig the trenches in the rocky soil at the preserve. We put in fence for over 30 lines, and I monitored the traps from March to November for the next 14 years, averaging about 150 snake records each year. Over that time, I recorded 14 species of snake, with southern copperheads *(Agkistrodon*

Drift fence at Sheff's Wood.

contortrix) being the most common in the uplands and cottonmouths *(Agkistrodon piscivorus)* in the bottomland. The rarest snake at Sheff's Wood was the glossy snake *(Arizona elegans),* and I only got one in 14 years. Trapping with drift fences often gives a very different perception of the community of snakes, compared to just flipping rocks or searching an area. For example, I was always told coral snakes *(Micrurus fulvius)* were very rare in East Texas, but in traps in the rocky area of Sheff's Wood, I often collected them after rain. Flathead snakes *(Tantilla gracilis)* were another example. I found them to be abundant when I flipped rocks but never got them in traps. They were too small to be captured that way.

I initially marked the snakes I trapped by clipping scales on the ventral part of the tail, just as Jesús had with his anacondas (Chapter 11). However, in 1989 a local fisheries biologist loaned me a pit tag reader, and I started implanting passive transponder tags (PIT tags) in the snakes. These tags were the size of a grain of rice, with a microchip containing an individual 10-digit number. I implanted them in the bodies of the snakes, and they lasted for many years. When I switched to them, I realized there were not as many cottonmouths at Sheff's Wood as I had thought. Instead, I was seeing the same ones repeatedly. One adult female cottonmouth even learned to come into my traps to feed on the frogs that were there. These repeated captures suggested to me that the local perception that cottonmouths were very abundant around ponds in East Texas was likely inaccurate, but rather, they were seeing the same individuals over and over.

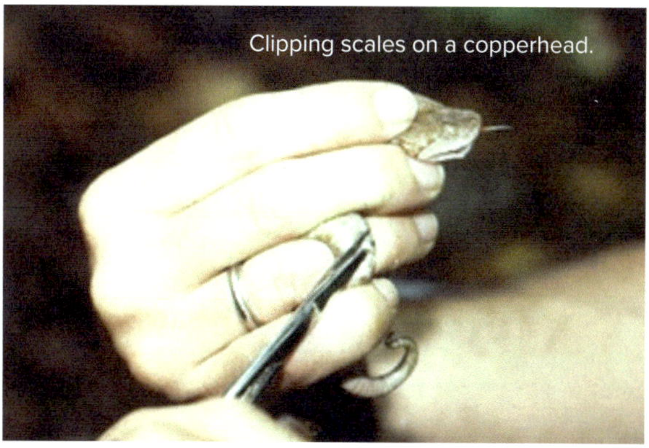
Clipping scales on a copperhead.

In 1998, I was funded by Texas Parks and Wildlife to do

baseline surveys of terrestrial vertebrates at a new wildlife management area that was being established on the floodplain of the Sabine River. It was an old hunting lease that was partially donated and partially purchased with Duck Stamp funds. It contained about 6,000 acres of bottomland hardwood and TPWD named it the Old Sabine Bottom Wildlife Management Area. I didn't know a lot about bottomland hardwood forests, except what I had heard conservationist Jim Neal talk about at environmental meetings. He always stressed how biologically diverse bottomlands were. I quickly learned how many species and numbers of animals were in such habitats. At Sheff's Wood, I typically got 150 snakes in my drift fences over the summer. At the Old Sabine Bottom WMA, I would get 150 snakes in a week!

Bottomlands contained a lot of woody debris that subsidized energy at the base of the trophic pyramid. For example, the Old Sabine in the winter flooded several times, and huge amounts of leaves, tree limbs and brush were brought down from upstream. The water spread out in the floodplain and the flow was greatly reduced, causing the organic material to drop out. I could actually walk in the flooded roads of the WMA. When the water moved back to the river, the organic matter remaining in pools became food for lots of invertebrates, which were then eaten by fish and salamander larva. When the water levels dropped in the pools, if the amphibians hadn't metamorphosed and the fish hadn't washed back to the river in later floods, they became prey for snakes, turtles, birds and mammals. The energy input in the floodplain was tremendous and was the reason this area had so many animals.

Neil in the floodplain at the OSBWMA.

With me concentrating on the reptiles and amphibians, I hired local naturalist Alan Byboth, who now manages the natural area at Sam Houston State University, to survey the birds and my graduate students to do the small mammals. I put out some drift fences but mostly used cover boards. They were much easier to put out and worked well in attracting amphibians and reptiles. I also ran turtle traps and minnow traps in the pools and drove the property many nights looking for snakes crossing the roads. During the day, Alan had routes on trails where he would stop and listen for birds. My students set Sherman live traps for native species of mice and rats throughout the winter.

However, 1998 was a dry year, so no floods occurred, the organic input was greatly reduced and the pools dried out early in the summer. We recorded much lower numbers of most of the animals. As I indicated in previous chapters, fieldwork is heavily influenced by the weather. Luckily, TPWD gave me a second year of funding in 1999, and with a different rain pattern, we did much better that year.

If people drove up or TPWD brought out visitors while we were checking traps or lifting coverboards at the Old Sabine Bottom, we always stopped and talked to them about our work and why baseline studies were important for wildlife management. In particular, we always explained why snakes as top predators were important. Everyone was interested in seeing snakes, even if the first words they spoke were about the ones they had killed. I know the parks and wildlife personnel appreciated hearing about our work, and if we convinced any of the public to not run over snakes crossing the roads in this WMA, it was worth any delay in our work.

It was also the reason I was willing to talk to newspaper and television reporters about some snake story in the news. Many people seemed to have a love-hate relationship with snakes. They loved hearing some story about snakes but it also caused a shiver down their spine.

One good example of how a student’s background knowledge helped me was François Brischoux, who came to my lab from France.

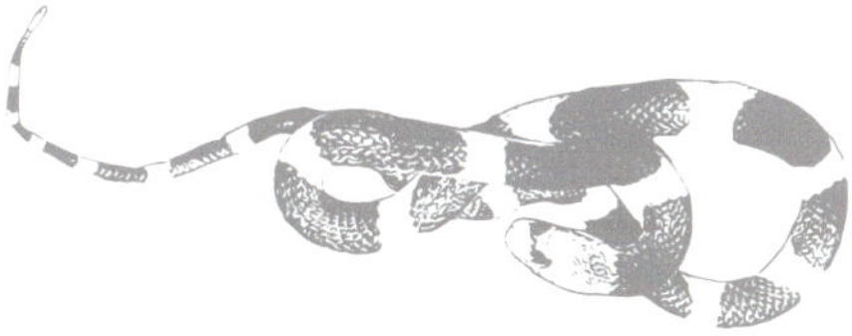

CHAPTER 15 | *What Students Bring to a Research Program: My Trip to French Guiana With French Herpetologists*

Training students is an important role for any scientists, including field biologists. Students in training work on projects that relate to the professor's area of expertise. Because of this, they increase the amount of research that a professor may be able to conduct. However, the ultimate goal is to teach the students how to be independent researchers so they can design and conduct projects of their own and run their own research program when they graduate. Only some of this scientific training is deliberate, rather much of what they learn occurs by the student conducting a project under the guidance of the major professor.

The best projects I found, i.e. the most successful, were ones that the student was excited to conduct. Primarily, this was because there are times when the work was difficult and the student had to push through those periods. It is also important that the project be tailored for the student such that the professor can successfully guide them to a successful conclusion. The advisor needs to be able to instruct the student in terms of what papers to read, what research aspects would be new and useful, how to design the project and even where to look for funds for the project. Therefore, it is critical that the student and professor interact successfully. Note, I did not say "well." Every student I have had has been different and most were fun to work with; but some were a struggle, although they still succeeded. However,

I did have a few quit for various reasons. My understanding of what the students were hoping for and their understanding of what they needed to do were both part of a successful research project.

A secondary function of taking students was what they could bring to my research program. Many of my graduate students had already conducted a project as an undergraduate, and many had some expertise or experiences that they could impart to my other students and my lab. One good example of how a student's background knowledge helped me was François Brischoux, who came to my lab from France. François was hoping to enter a graduate program in the Centre d'Etudes Biologiques de Chizé and would get a tuition reduction for each scientific paper published with him as an author. Olivier Lourdais of the French Centre of Biological studies knew of my research and recommended to François that he work with me. So François came and spent a summer working on a project studying reproduction in cottonmouths at the Old Sabine Bottom Wildlife Management Area. We did publish that data, and François got into the graduate program there and is now a professor. However, for me, an important part of his coming to my lab was the connection I made with French herpetologists and, with that, the opportunity to go to French Guiana.

I was still interested in trying to find specimens of the South American water snake, *Helicops angulatus,* that I thought might be able to switch from laying eggs to having live young. François told me that they occurred in French Guiana on the east coast of South America. In talking to him, I got excited about the idea of taking a field trip there. French Guiana was one of three countries that were colonies I had learned about way back in my high school geography classes. French Guiana is still a colony of France as its coast is a French tourist attraction and France has its space facilities there. British Guiana became independent in 1966 and is now Guyana. Dutch Guiana became Suriname in 1975.

Oli was interested in working in French Guiana, and as it was still a French colony, he could get permits directly through the wildlife offices in France.

He also knew several local herpetologists who were conducting herpetological surveys there. So we made arrangements for us to go to French Guiana.

Therefore, on January 1 of 2003, I headed to Cayenne, French Guiana (northeast side of South America) to meet Olivier Lourdais and François Brischoux to look for *Helicops angulatus.* They picked me up at the airport, and we headed to the nearby town of Montjoly and the home of Mael Dewynter, which would be our base for the next 10 days. French Guiana has a very low population density with most people in the country living along the Atlantic coastline. The interior is nearly all forested, and most areas have some measure of environmental protection as it is an Amazonian Park and has restrictions on logging. Several local biologists (Jean-Pierre Vacher and Philippe Macquet) were doing baseline studies of the amphibians and reptiles, so we were going along with them on their surveys. I rented a car so we would be able to take different routes wherever we hunted.

We all stayed in Mael's house, so it was crowded and most of the conversations were in French. They were all younger than me and definitely enjoying the adventure. We spent most of the 10 nights I was there driving different protected areas. The first few nights, we drove a road that went up

The Kaw Mountains.

Neil in French Guiana.

into to a 2,400-hectare forested reserve called "Tresor," in the Kaw Mountains. The first night, we didn't get any snakes but we saw lots of tropical frog species I had never seen before, *Scinax ruber, Pristimantis marmoratus* and *Leptodactylus mystaceus.* The coolest was a *Pipa pipa* we found in a pool. It was a female carrying eggs, embedded in its back.

Tresor was relatively undisturbed habitat with high biodiversity. The caretaker at the entrance to the reserve told us some poachers had even killed a jaguar recently. The second night, we met up with another car containing Vanessa Hequet from the New York Botanical Garden and Daniela Grabul, a herpetology student from Austria who was studying dendrobatid frogs (poison dart frogs). They gave us a head start at dusk to double our chances of finding animals crossing the road.

Our group quickly found a 2 ½-foot snake called a brown sipo, *Chironius fuscus.* It was skinny and brown and looked harmless, but as I was unfamiliar with any of the snake species there, I waited until Jean-Pierre informed me it was nonvenomous before I grabbed it. We also stopped for some huge worms we thought were snakes. Search images take a while to develop when you are night hunting. Then a fer-de-lance *(Botrops atrox)* crossed in front

Mael Dewynter and Phillippe Macquet measuring snake. (Courtesy of Phillippe Macquet)

of us. I did recognize this venomous snake and grabbed it out of the grass on the side of the road, using my tongs. This species caused more bites in the tropical Americas than any other viper. Oli pinned it and grabbed it behind the head to measure it. All of them were more cavalier about this than I would have been. Maybe it is an age thing as I do remember pinning lots of venomous snakes when I was younger. Of course it could be because I was bitten twice? I learned from my second graduate student, Jan Perry-Richardson, that getting the snake to crawl into a clear tube was a much safer way to measure them. I had a set of all sizes but didn't bring them on this trip. Next, we got a DOR vine snake, *Oxybelis aeneus,* a little further up the road. DOR is the standard abbreviation herpetologists use for "dead on the road" since a lot of the animals found are run over.

The highway dead-ended in the Kaw Mountains, where there was actually an inn just for scientists. As usual, we were underfunded and slept in the car park there while waiting for the other car to show up. I walked down a trail while we waited, but it was so dark that, if my light had gone out, I would have never found my way back! When I got back to the car, Jean-Pierre warned me not to wander down trails at night by myself!

The girls' car finally showed up and they had caught an *Imantodes* (blunt-head tree snake) and a *Corallus* (tree boa). On the way back, we saw a DOR *Chironius* and then stopped at a roadside ditch, where Mael caught a small anaconda. I saw a second one that I at first thought was a *Helicops.* Then I griped, "All I catch is anacondas." Of course, the others were very happy to get an anaconda as they weren't very common. Mael then caught a snake he initially thought was the harmless *Helicops leopardinus* but then wondered if it wasn't a *Micrurus* (coral snake). Since it had just bitten him, we hoped it was a *Hydrops,* which was a harmless coral snake mimic called a water coral. Apparently it was, since he didn't die. We got back to the house at 2 a.m.

The next day, we drove back to the reserve to photograph the snakes we caught the previous night. Philippe was filming a television program for children about animals in the canopy. I also brought a video camera and used it a lot. It is fun to look back at the video, but in putting this book together, I regretted I didn't have photographs of the trip. While we were in the forest photographing the animals, I became very dizzy and had to sit down. I took a Dramamine and laid down on the forest floor but didn't get better. I threw up several times and then felt well enough to at least drive back. I remembered then that I had the same problem in Haiti with the anti-malaria drugs. I stopped taking them! Surely malaria can't feel this bad.

After lunch, which I didn't eat, we headed out on a road to the town of Guatemala. The landscape quickly turned into savanna (grassland with a few trees), and we stopped and hunted around a couple of shallow pools. Mael found a *Erythrolamprus cobella,* the same spotted water snake I had found in Trinidad. He pointed out some of the buildings of the Spaceport on a hill. The French sent off their satellites from French Guiana. With fire ants and mosquitos competing as to which could bite us the most, we didn't stay at this site long. We stopped at another pool, near a forest edge, and got a *Leptodeira* (cat-eyed snake). We also found a three-toed sloth that was coming down from the trees. Then we met up with Philippe's parked car and

A three-toed sloth. (Courtesy of Phillippe Macquet)

saw him searching on the side of the road. He had a *Pseudoboa,* a brilliant red-orange snake. We headed back to the house, and when Mael showed up, he had caught a 6-foot redtail boa. He had it in a plastic bag, so I gave him a cloth snake bag to put it in. He also brought back several DOR snakes to preserve, including *Erythrolamprus miliaris, C. hortulanus* and *E. cenchria.* Mael seemed to be the best of us at finding snakes. We then drove into Cayenne to send off emails at a cyber café and to buy hardware cloth to make snake traps, since I had not brought minnow traps this time.

After some French "inertia," we took the snake traps we built and placed them near the Mahury River, in a swampy area that had shallow canals that looked perfect for *Helicops.* Mael had arranged for us to collect in an abandoned brick factory on the site. Then we went night driving up Kaw Mountain again. We got a live parrot snake, *Leptophis ahaetulla,* a DOR *Chironius fuscus* and a little ground snake, *Atractus,* but I don't know which species it was. We turned around at the end of the road, and on the way back, we got a caecilian, *Rhinatrema bivittatum,* a tropical limbless amphibian that had scales in the skin (see Chapter 2).

A coral snake mimic *(Erytholamprus aesculapii).* (Courtesy of Phillippe Macquet)

We released the anacondas in the Mahury river and drove up toward the reserve again. Mael and Philippe caught a coral snake mimic, *Erythrolamprus aesculapii,* that was bright orange and black, and as we handled it, it regurgitated an *Atractus.* We heard a troop of golden marmosets in the trees where we stopped. It was hard to see them, but I did get some video of them and they were definitely little monkeys. We got another *Chironius* and a water snake, *Erythrolamprus typhlus,* on the way back. When Mael's group came in, they had caught a dark red and black tropical flat snake, *Siphlophis compressus,* and a couple more *Chironius.*

We went up to the wildlife office near Mont Joly. The agent there, Philippe Gaucher, brought out beer and snacks and showed us some frogs, *Trachycephalus hadroceps,* explaining that the female laid infertile eggs in the water-filled cups of bromeliads for its tadpoles to eat. He also had a *Neusticurus,* an aquatic lizard that the *Helicops* was known to eat, as I had read. By the way, when I visited wildlife offices in the states, they never offered me beer!

During afternoon breaks back at the house, François and I worked on our cottonmouth paper and managed to finish it before I left. We also watched some movies. "Family Man" was really boring in French. One day, we went over to a tree-growing facility called Silvolab, in the commune of Kourou, and heard talks by Peter Narins from UCLA and Walter Hödl from Vienna. They were bioacoustical researchers working on a "Robofrog" at a field station called Aratai, three hours into the forest. They explained that one had to get permission from the local indigenous people before going into the interior of French Guiana.

Several afternoons, Oli, François, Jean-Pierre and I checked the traps we had put out. We caught some of the water snakes, *Helicops leopardinus,* each time. This species is related to the ones I got in Trinidad but was banded and had a beautiful red and black belly.

We went to buy groceries for a longer trip to Iracoubo, where the new hydroelectric dam was planned. We picked up some sausage covered with white mold, red deer pate, bananas and canned applesauce. A bit of a strange picnic lunch in my view, so I got some peanut butter crackers just to be safe. The guys explained that grocery stores here were called "Chinese shops" because all of them were owned by Chinese. We went by the Space Center on the way, and Mael said the launches could be felt in Cayenne. We stopped in Sinnamary for the bread we had forgotten. The guys were not very organized, and I was becoming aware of that.

We stopped at a Nature Center on the way and climbed a tower there to eat the sausage with white mold. It was actually fine, just looked bad. We went down some nature trails while waiting for the others to arrive. We saw yellow-headed vultures, a great white heron and a small kingfisher. We drove on to the savanna at Poissons and took a little dirt road into the grassland. We stopped at a couple pools on the jungle periphery and saw depressions that had been dug out next to the pond by the frog, *Boana boans.*

They were full of her tadpoles. The little pools protected them from fish predators until the rains reconnected the depression to the larger pond.

As we walked around the pond, Oli saw a 9-foot anaconda and grabbed its tail to pull it onto the land. We took lots of pictures. After hunting in ditches in the savanna for *Helicops* for a while, we headed on toward the Petit Saut Dam. As we drove on that night, we got two little brown snakes, *Atractus badius,* and a water snake, *Erythrolamprus typhus.* That night, we set up camp in a pole barn that was meant for public camping on the route de Petit Saut Dam. It was a very nice facility with hooks for hammocks. Next to it was a beautiful forested stream that we explored.

We ate another meal of deer pate and applesauce. I was glad I also had my "magic crackers," as the guys called my peanut butter crackers. We drove the road for a while after "dinner" but only got a huge waxy-monkey treefrog, *Phyllomedusa bicolor.* Any time we walked off the highway into the woods, the mosquitoes drove us back into the car pretty quick. The guys slept in hammocks that night. I blew up my inflatable pad, so I was comfortable — although it actually got cold. Before we hit the sack, I gave the guys lessons on Texas sayings like "happy as flies on a dead pig in the sunshine."

The next morning, I really missed my coffee! Mael and I were up first and took the car to go birding up and down the highway. We saw more golden-handed tamarin monkeys in the forest, although we mostly just saw the vegetation moving. We saw a couple of white hawks and some parrots but little else. We did collect a young fer-de-lance, *Botrops atrox.* It still had a yellow tail that the juveniles used to lure prey by waving it like a worm. We also found a coral snake, *Micrurus psyches.* It was black and white, without the red bands that our corals have. It was so small that Mael picked it up with a plastic bag. I got a headache due to the lack of caffeine! Mael said they would have some at the biologist office but it wouldn't be very good.

On the morning drive to the dam, we got a vine snake, *Oxybelis anaeus,* and another DOR *Erythrolamprus thyphus.* The biologists at the dam told us they were working on building a museum that would include displays of the local animals. There was a big program in 1995 to relocate all the animals when the dam was filling. Mael was right about the coffee, it was awful! We headed back to Mont Joly and into Cayenne to return my rental car.

Mael and I checked the aquatic traps one last time and we got three more *Helicops.* All have differently colored bellies. They bite a lot more than the *Helicops angulatus* that I had collected in Trinidad, and the bite hurts for such little guys.

I took Air France home with stops in Trinidad, Martinque and Haiti this time, but with three nice meals. At my hotel in Miami, I called Julian Lee at the University of Miami and arranged to have breakfast with him in the morning, before my flight to Tyler. I had known Julian back at the University of Kansas when I was an undergraduate. That night at the hotel, I slept on a real bed! This trip might not have been so successful in terms of getting animals for my research, but it sure was fun. Thanks, François.

It was interesting to hear about all the exciting work Anne and her students were doing.

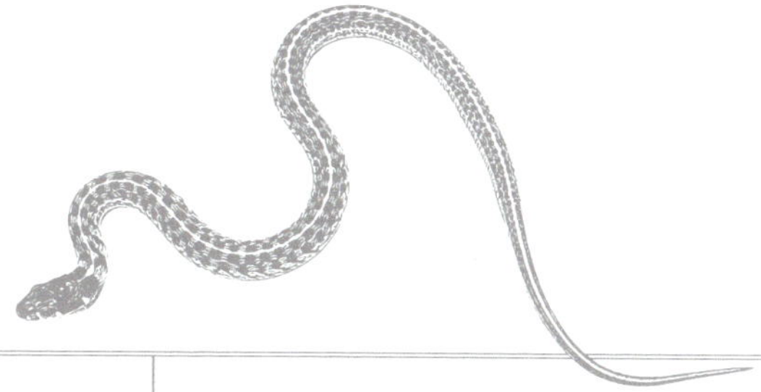

CHAPTER 16

Increasing Diversity in Field Herpetology: Snake Camp at Eagle Lake With Dr. Anne Bronikowski

Although research in field biology can often be conducted with little or no funding, having grants allows a scientist to do much more in less time. In particular, being able to pay students doing the research means the students can spend more time on the projects. Therefore, I always paid attention to any opportunities to apply for research funds.

In 2009, I ran across a National Science Foundation program (NSF) called the Research Opportunity Award (ROA). It was designed for professors in smaller schools to collaborate with professors in schools who already had NSF funding, with the idea of helping them develop their own NSF-funded research by learning from the funded professor. I had been aware of Dr. Anne Bronikowski's research on garter snakes for a number of years and knew she had NSF funding, so I asked her if she would be interested in working with me on a ROA project. She checked with the NSF director in charge of her research, and he agreed that we should apply.

I received two of these ROA grants over three years and used them to fund master's students in my lab to look at the role of energy intake on Insulin Growth Factor I and how it controlled age at first reproduction. These projects were not field projects per se but had important field applications in how quickly animals grew up, reproduced and grew old (i.e. the pace of life). My interaction with Anne and her students got me interested in visiting her field site at Eagle Lake, California, where she was studying aging in garter snakes. This wildlife recreation area was a long-term field site in which garter snakes had been studied for over 25 years. Stephan Arnold at Oregon State University, who had been Anne's advisor, had started working with garter snakes at Eagle Lake in the early 1970s. Steve's work was primarily looking at the evolution of prey preference in wandering garter snakes. One population ate slugs, whereas another ate fish and frogs. Anne was interested in the role of growth, aging and senescence in these different populations. Anne worked there in the summers and, like me, also brought back animals to her lab at Iowa State University for further study. Particularly, she looked at the number and size of the offspring of the females from the different localities. With the ROA grant, she sent her students down to Tyler to help my students learn how to draw blood and analyze the IGF1 hormone in the snakes in my project. These were African house snakes that we could raise up to breeding in less than a year. We placed them on high and low energy diets and then looked at their reproductive differences and their hormone levels. We then collaborated on writing up the results (Sparkman et al., 2010; Byers et al., 2010).

I wanted to see where Anne was working, so on June 30, 2009, I headed to their "Snake Camp" at Eagle Lake, California. I had visited the site with Rich Seigel when we were working on our garter snake book back in 1995 (Chapter 10). Eagle Lake is a California recreation area that had two ecotypes of wandering garter snakes *(Thamnophis elegans).* Ecotype refers to the fact that the snakes are the same species, but because of the habitat they live in, each population has evolved different physiological and ecological traits. Eagle Lake was particularly important to Anne's work because the two ecotypes of these garter snakes occurred very close to each other, but had very different life-history strategies. Life history refers to traits that organisms have, such as how many offspring they have, how big their babies are, how fast they grow and how often they reproduce.

One ecotype of the wandering garter snakes at Eagle Lake lived in meadows above the lake with few predators and ate worms and frogs, so they grew slowly and took several years to mature. They also had only a few large babies every other year. Whereas the snakes around the lake ate fish and had many small offspring that grew up fast. This strategy had likely evolved because the babies were prey for eagles and other birds. Anne and her students were looking at the physiological controls of this "pace of life" by examining in the two ecotypes the hormones that were likely involved in controlling the growth rates (Insulin Growth Factor 1). I was good at catching garter snakes, not drawing blood, so I was mostly a field hand.

I flew out of Tyler's small airport, which normally would have been easy, but for some reason, there was a line at security. I had forgotten to put my pocketknife in my checked luggage and there was no time to move it, so security confiscated it. I liked that knife! I flew into Reno, but there was no one at the airport baggage claim to meet me. I had neglected to bring a phone number for Anne with me. I called my office phone and

Neil catching garter snakes at a meadow site.

checked messages, and then found a wi-fi area and tried to see if she had sent a message to my email. Nothing! I should have had a backup plan! I had no choice but to wait. Finally, I heard an announcement that I thought was asking about me. Yeah! It was Tonia Schwartz, one of Anne's grad students. Apparently, Fred Janzen (Anne's husband who is a turtle biologist) underestimated the time to drive to Reno from Eagle Lake. By the time they got to the airport, the Dallas arrival sign was off the baggage carousel screen, so Tonia didn't know where to look for me. She finally thought to page me.

On the way to Eagle Lake, we stopped in Susanville to pick up liquid nitrogen to flash freeze tissues for the DNA work. Anne wanted to know how closely related the individual snakes were in each population. We also stopped at a grocery store — for beer and liquor, of course. When we got into the camp, everything was already set up. There were about a dozen tents around, with about 20 people already working at tables. Anne had a program at Iowa State University where she took female undergraduate students into her lab with the

Steve Arnold showing students how to preserve snakes found dead on the road.

goal of increasing diversity in science. NSF grants had started requiring efforts to increase diversity (both minorities and females) when they funded projects.

Lynn Houck, a salamander biologist from Oregon State (Steve Arnold's wife), showed me where all the ice chests were with the food, where the pay showers were and how to get tokens to use them. Cold water only. Steve was showing several students how to preserve a gopher snake and pacific rattlesnake that they had found dead on the road the night before. The specimens had been hardening in formalin overnight, and they moved them over into 70% ethanol.

I just stored my gear, and Anne took me out to see the lake population of garter snakes. The snakes were crawling around the large rocks along the edge and foraging for fish at the water's edge. They were hard to catch among the rocks, but we caught several young and two of those had fat bellies, so we palped out what they had eaten by pushing the lumps up to the mouth. They each threw up several minnows. One snake was all black, which is termed "melanistic." Their dark color allowed them to warm up faster but

A wandering garter snake site at Eagle Lake.

also made them more vulnerable to predators, as they were easy to see on the gray rocks. It got dark as we were hunting, and we got a bit lost on the way back but finally made it back for supper.

That night Hartwood Welsh, of the U. S. Forest Service, visited camp. Hart had shown Rick Seigel and me around his sites for Couch's garter snake when we came to California in 1995 (Chapter 10). Dr. Al Bennett, a biologist from Berkley, was also staying for a couple days. This trip was kind of "old home" week for snake biologists. We sat around the fire and talked about our research projects as well as other field trips people had taken. Steve and I told the students stories about catching snakes and crocodiles in Australia. It was cold that night and my sleeping bag wasn't warm enough, so the next day, I borrowed one of the extra bags.

The next morning, I woke first and went to make coffee, but the coffee was not in the bin labeled "coffee." When Al woke, he showed me that the coffee was in the bin labeled "cheese." He says this was typical for Snake Camp. Field biologists aren't always great organizers.

It was cloudy that morning and the snakes were not out yet, so we ate a leisurely breakfast before we went up to the meadow sites. One meadow site was named "Papoose" and was a picturesque old farm in a flat grassy area, between rocky outcrops. It was beautiful! But wet. "Nameless" was another meadow we went to. It was not on any map. Hence its name. It was a meadow with granite outcrops on the side and pools in the middle with just a little water. Some small tree frogs *(Pseudacris regilla)* were hopping in the grass, and the snakes were hunting them. The students and I started catching garter snakes. We had to watch for their movement and then pounce! Whenever we caught one, we quickly brought it to Tonia to draw blood. She had to do this within just a few minutes before the stress hormone, corticosterone, could be released into the blood. She was looking for Insulin Growth Factor I, which she believed might be the hormone controlling the differences in growth rates between the two ecotypes.

Visiting a meadow site during Snake Camp.

We got 22 *Thamnophis elegans* and a couple of red-sided garter snakes, *T. sirtalis.* Tonia drew blood from all the *T. elegans*. Afterwards, we climbed up the rocks and looked at an old Indian fort on the hillsides. Some of the students looked for obsidian rocks.

We relaxed around the fire that night, and I slept better since I put a second sleeping bag over mine. The next morning, Steve took me around to some other lake sites that he had hunted in the past. Some of his old hunting areas had been converted to private homes, so it was hard to find many places to look. We got permission from one landowner when Steve told him what we were doing. He explained how the research related to human aging. However, we only found a few snakes there and just measured them and let them go.

The hunting wound down in a couple days, and Anne drove me back to the airport. I enjoyed this trip mostly because I got to catch garter snakes again. I didn't come back with any great new ideas for projects, but that didn't happen on every field trip and I had enough ideas to keep me busy anyway. It was interesting to hear about all the exciting work Anne and her students were doing on aging and life history traits. It was also nice to see how many females were interested in herpetology now, versus when I got started.

Even though communication methods among the societies' members have changed dramatically with the advent of email, these meetings are still critical for biologists.

CHAPTER 17

Conclusion: Scientific Meetings on Four Continents and an Island

When I started my career, the number of herpetologists in the world was only in the several hundreds and communication among those scientists was through letters and at scientific meetings. These scientific meetings were held by societies that, in general, had been organized for presentations and communication around specific scientific disciplines like herpetology. At meetings, scientists gave talks about their current research and also interacted informally during breaks and in the evenings. Some societies met annually like the Herpetologists League and the American Society of Ichthyologists and Herpetologists, whereas others like the World Congress of Herpetology and the Snake Ecology Group only met every four or five years. Some meetings, like the Biology of the Vipers conference held in Sweden in 2000, were organized around a specific theme and only met once.

Even though communication methods among the societies' members have changed dramatically with the advent of email, these meetings are still critical for biologists to learn about the work that other herpetologists are doing. Publication of results of scientific studies requires peer-review, so new research can take months or even years to be available for other scientists to read. The information and ideas presented at meetings can let other herpetologists know about more current research, stimulate interactions like mine with Rich Seigel and thus lead to much faster advances in our knowledge. In this final chapter, I will give some examples of those meetings and how they affected my research.

I started going to herpetology meetings when I was an undergraduate at the University of Kansas. Although there were societies around the world, the biggest were those in the U.S. and they typically had a few hundred attendees, almost all Caucasian males. The First World Congress of Herpetology was held in 1989 in Canterbury, the United Kingdom at the University of Kent. It was the largest meeting of herpetologists at the time with over 1,300 attendees from all over the world. It sparked additional World Congresses, which have been held about every five years. I attended that first meeting and met colleagues that I had only known through their publications. We stayed in dormitories at the University of Kent, and although the talks, particularly those from researchers from other countries, were very interesting to me, the most useful parts of the meeting were the evenings when we gathered at pubs in the dormitories to talk about our research.

For example, during a nocturnal congregation of biologists who were working with amphibians, a recurrent theme discussed was how their individual study animals were disappearing, even in protected nature preserves and refuges. These extinctions were so prevalent and so alarmed the biologists at this meeting that the group decided to form a worldwide research effort. The Amphibian Decline Taskforce was established to try to find out what was going on with frogs and salamander populations. Since that time, the group has funded many conservation projects around the world and identified many of the environmental issues affecting amphibians. This might never have happened without this meeting!

Although I learned a lot interacting with other herpetologists at this first world congress, the organizers had only planned field trips to castles and historical venues, not to look for amphibians and reptiles. Later congresses did arrange field trips, and I'm going to describe ones that I took during the Second World Congress of Herpetology, held in Adelaide, Australia, in 1993.

The Second World Congress of Herpetology organizers arranged optional field trips for attendees prior to the meeting, and a group of us studying the ecology of snakes was invited to meet after the congress. My wife, Joan, and son, David, flew with me into Adelaide by way of Sydney before the meeting. We had arranged to take a couple of guided trips. The first was up to Arkaroola Village, in an outback region called Flinder's Range. Dr. Gordon Burghardt of the University of Tennessee, who studied snake behavior, and his wife and their two girls, Leana and Karin, joined us on the trip. It was a 10-hour drive through barren land, but we were comfortable in a minivan. However, David did get car sick in the very back, probably because of fumes from the extra gasoline stored for the long trip through the outback, where gas stations were limited.

The Arkaroola Village in Australia.

Arkaroola Village was a sheep ranch that had been converted to a wildlife sanctuary. We saw lots of birds and mammals, like the rock wallabies we watched hopping from rock to rock at Bolla Bollana Spring. But Gordon and I were more excited to catch lizards, like a red barred lizard and the 12-inch shingle-back skink and the blue-tongue skink on the trail during the 7-km walk back to our lodge. After a lunch of kangaroo meat pie and veggie pasties, Doug Sprigg (the son of the founder) took us in a jeep around the rough and steep terrain to various amazing and beautiful sites like Barearanna Gorge, Jasper Twins, Stubbs Waterhole and the Ochia wall. The latter was made of a red and orange rock that was used by the aboriginal people to make paint to draw on rock walls. I had read old "Napoleon Boneparte" mysteries from the '20s and '30s, so it was interesting to me to see the "outback" described in these books.

A blue-tongue skink.

The next day, we took the Mawson Valley trail and saw lots of euros (as the locals called the rock wallabies). We lunched at Radium Creek, where we saw wild camels that were descendants of those that had been used to haul heavy loads in the desert. On December 25, we had a huge Christmas feast with the owners, Reg and Guiselda Sprigg; and on Christmas night, we climbed Mt. Guiselda and looked at stars including the Southern Cross, using a telescope at their observatory. The next day, we started the long trip back to Adelaide.

From Adelaide, we took a second field trip over to Kangaroo Island. We flew over on a five-seater plane and Joan got to be the copilot. On the island, we went in a bus full of other herpetologists on a fast tour of the island. At Pennington Bay, the guide showed us lots of unusual plants, like maller trees, pussy cat tail grass and bottle brush, and also huge termite mounds. He also gave us the history of Cape Willoughby Lighthouse, the oldest in South Australia, as well as Devils Kitchen with its huge red rocks and Point Scapes, where many shipwrecks had occurred. However, on most stops, the herpetologists would immediately head out looking for reptiles instead of paying attention to the guide or the time we were told to be back at the bus, so the guide was a bit frustrated.

At our hotel after dark, Joan, David and I walked around and found a "herd" of 25 or so wallabies. The next morning, the hostess told us where to go to see the rare glossy black cockatoo, and we found them! The bus ride that day took us to Seal Bay, where we saw about 50 endangered Australian sea lions. This area was only one of two places in the world where one could walk among them. The other was the Galapagos. We then went to Kelly Hill Caves and Flinders Chase National Park. At the homestead in the park, we saw koalas, kangaroos, emus and wallabies. We also saw fur seals at the nearby Cape du Couedic. After a long day, we took a ferry back from the island and then a bus back to our hotel in Adelaide. We had seen only a few snakes and lizards, but it was great to see all sorts of Australian birds and mammals.

Australian Sea Lions at Sea Bay, Kangaroo Island.

The next day, I registered for the meeting and then we went to Cleland Wildlife Park, where we saw hairy-nosed wombats and a Tasmanian devil. I always taught my Vertebrate Natural History students about "convergent evolution" using Australian mammals as examples, so it was great to actually see them.

Then I met up with Bob Kaplan from Reed College, who studied amphibian life histories much like I did with snakes. We attended the meeting together for the next couple days, while Joan, Linda and Jessica Kaplan and David went exploring. After the conference was over, Joan and David flew to Sydney for their flight home, and I took a flight up to Darwin in the "Top End" of Australia. This Northwest territory was so huge and sparsely populated that it was not even a state.

A group of about 15 of us studying snake ecology were invited up to Tom Madsen and Rick Shine's field site at Fog Dam. We stayed at Middle Point School near the Coastal Plains Research Lab. It had an open-air dining area where we ate and had our discussions. We slept in sleeping bags on the floor in the one-room classroom. During the day, the grad students told us about their research. One student was working with death adders *(Acanthophis antarcticus)* which are heavier bodied than most Australian snakes. I showed him the short tongs that I used with copperheads and suggested they would work well. He was not impressed. The Australians just used hooks to control their elapid snakes while holding on to the tail! A copperhead would bite you if you did that!

The first day, while we were eating lunch, a 6-foot saltwater crocodile *(Crocodylus porosus)* wandered through the front yard and 15 herpetologists all ran and jumped it. That croc chose the wrong place to walk! It turned out he was an escapee from a neighboring crocodile farm. That day, we took a trip to Arnhem, which was aboriginal land, stopping first at Jumping Crocs in the town of Humpty Doo. On the Jumping Crocs tour, we went on the Adelaide river in a boat with high glass walls. All along the river banks were

A saltwater crocodile.

large saltwater crocodiles. The boat guide would slap the water, and all of the crocs would come over to the boat to feed on sheep's heads, which dangled on rope from the upper deck. They would jump up in front of us, almost completely out of the water to get them! Some of them were over 12 feet in length. Incredibly strong animals, it reminded me of riding in the SREL van with the 11-foot alligator back in South Carolina (Chapter 7).

On the way to Arnhem, we stopped for a perentie *(Varanus panoptes),* a 3-foot monitor lizard, in the road and chased it down. Because of the lack of placental mammals in Australia, the lizards had evolved into the large dominant predators. In Kanadu, we looked at aboriginal cave paintings and went looking for the rare Oenpelli python, *Morelia oenpelliensis.* The tribe that had lived on this land were all gone, so no one lived in this huge area. We didn't find any snakes, but as it grew dark, we hunted for gekkos among the rocks. We had flashlights, but walking around alone in the dark among the large boulders was very confusing, so I did not go far from the bus. Getting lost there would have been a real problem.

The next day, we split up the group and took two buses for a day trip around the area. Matts Olsson (University of Gothenburg), Steve Arnold

Rick Shine and Neil with frilled lizard.

(Oregon State University), Whit Gibbons (Savannah River Ecology Laboratory) and Rick Shine (University of Sydney) were in my bus. We stopped for several frilled lizards we saw on trees. Even though they were large, like our fence lizards, they would hide behind the tree and look around the trunk at us. If you slapped your hand around the tree without moving your body, you could catch them. Steve Arnold got the first one, a female. The next one was a large male, and my turn. Just as I was about to slap it, Steve yelled "Watch out for the fangs." So I hesitated just a second and missed it as it ran up the tree! Steve felt bad, so he shook the little tree until the lizard fell out. Right into my hands. It opened its mouth at me and it did have these enlarged teeth that would have hurt!

During the next few days, we got several snakes including an eastern brown snake *(Pseudonaja textilis),* which was so dangerous the local herpetologists iced it down to photograph! We also got a bandy-bandy *(Vermicella annulate),* which was black and white like our eastern king snake, and a golden tree snake *(Dendrelaphis punctulatus),* one of the few nonvenomous snakes in Australia. I learned from the locals that the Australian snakes, although most were venomous, varied a lot in how dangerous they were to humans.

One night, we went over to the mangroves along the Darwin River to look at saltwater crocs. As we saw a smaller one, Steve and Whit jumped out of the boat and chased it down. Whit obviously was experienced with catching alligators back in South Carolina, and Steve was just competitive. They both missed it. We did catch two species of mangrove snakes *(Myron richardsonii* and *Fordonia),* both of which ate crabs. There were also cute little mud skippers (like a little catfish) that were hopping over the mud and down holes when we got close.

The next day, we hunted at Fog Dam near the research lab and caught keelback snakes *(Tropidonophis mairii),* water pythons *(Liasis fuscus),* several species of frogs *(Litoria caerulea and Litoria dahli)* and a snake-necked turtle *(Emydura oblonga).* Again, it was special to me to see examples of animals that I discussed in my classes back in Tyler. That night, we headed to the swamps in Jabiru to hunt for *Acrochordus*, the file snake. These snakes were quite strange, with skin like sandpaper. Steve caught one as he stepped out of the bus. So I was determined to catch one also and went off in a different direction by myself. As I walked around the edge of the dark water, I heard someone shouting to me that there was a croc heading my way. I thought it was Steve just kidding me again, until I saw the croc's eye shine with my flashlight. There had just been a story in a local newspaper about a crocodile killing a swimmer, so I decided to go back to join the group! This was the second time in my life that I felt like a prey item. The first was while hunting anacondas in Venezuela (Chapter 11).

A file snake.

Since that second meeting, there have been seven more World Congresses of Herpetology around the world. However, the number of scientific societies had expanded dramatically and many more annual meetings were occurring. The only other world congress that I had the time and funds to attend was held in Stellenbosch, South Africa, in 2005. We did explore before the meeting but mostly looked at African mammals, not snakes that time.

One organization, the Snake Ecology Group, was started specifically for those herpetologists studying snake ecology and behavior. This loosely organized group was formed somewhat by accident from a symposium that Pat Gregory (University of Victoria) and I organized in 1986 in conjunction with the annual meeting of the American Society of Ichthyologists and Herpetologists in Victoria, British Columbia. This symposium had 23 talks focused on the behavior and ecology of snakes, given by 35 presenters from around the world. It was so successful that, afterward, we formed this loosely organized group that met about every five years.

The first meeting of the Snake Ecology Group was organized by Rich Seigel at Southeastern Louisiana University's field station on Lake Ponchartrain and had about 40 attendees. I organized the next meeting at Camp Tyler in Tyler, with over 120 herpetologists from universities and zoos attending. This one was so successful because, with the advent of email, I was able to send out repeated announcements. I convinced a couple of the most well-known herpetologists to commit early, and by mentioning their attendance in my emails, I convinced others to come. It was an outstanding meeting and not just because of the southern cooking.

The Kunigami Environmental Education Center in Okinawa.

The Snake Ecology Group has met at various field stations several times since. The sixth meeting was organized by Dr. Akira Moira of Kyoto University and was held in 2013 in Okinawa, Japan, at the Kunigami

Environmental Education Center on the isolated north end of the island. It was a great facility with both dorms and small apartments.

Around the nature center was the Yambaru Discovery Forest, which we explored before breakfast each day and found Anderson's alligator newts, *Echinotriton andersoni,* on the wet trail. During the day, I went to talks and Joan explored River Song Trail and other paths through the forest. I gave the last talk of the first day on snakes of the U.S., but everyone was tired, so only a few questions were asked. We then had an early dinner because of the scheduled nighttime field trips.

An Anderson's alligator newt.

Joan and I signed up for a streamside night hike and headed out with a dozen or so of us in a small van. On the drive, we stopped for a hemihabu *(Oviphis)* crossing the road. It apparently was not too dangerous as everyone photographed it up close. The main hike was up a narrow forest stream, and we saw three species of frogs. Tom Madsen caught an *Amphiesma,* which was striped similar to our garter snakes. I had kept this snake species in captivity years earlier, so it was exciting to see one in the wild. I then caught a red banded snake, *Dinodon,* which bit me repeatedly. These snakes have very sharp teeth to hold on to frogs, so I bled a lot.

The next morning, along with the normal breakfast items, we got a slimy, pickled algae? Akira had arranged for us to have traditional Okinawan food each meal, and I at least tried them all. The talks this last morning were about snakes of Japan, Taiwan and Australia and very interesting. The bus then took us on the winding ride back to the city.

After a visit to the Churaumi Aquarium, the second largest in the world at the Okinawa Marine Park, Joan and I had arranged several diving trips.

Photographing a hemihabu in the road.

It was quite a different experience from our diving trips in the Caribbean. We were responsible to get all our own gear together each morning. We then picked up lunch at a convenience store for later and went to the boatyard to take a ride to our dive spots. Our trip the first day was to the outer islands. There were lots of fish there, but the habitat had been damaged a lot as the reefs were not protected. We did see several sea snakes, *Laticauda colubrine,* hunting among the rocks for fish eggs.

The next day was a rough ride out to the islands but worth it, as we saw sea turtles. Most of the coral there was dead also. There were no regulations about divers not touching it. Still, we saw lots of fish. We went through an underwater cave and into a large chamber, but lots of people were there and their fins had stirred up the sand, so we couldn't see much.

The last day, because the waves were even rougher, we took a smaller boat and went to a nicer spot much closer to the mainland. We saw five species of anemone and the different clown fish associated with each of them. We also saw eels, spiny toadfish, octopus and black scorpion fish. It was always great to be able to add some fun field time when I went to a scientific meeting!

The Biology of the Vipers meeting held in Uppsala, Sweden, in May of 2000 was a very different event in that it was a stand-alone meeting. The speakers were specifically invited to present their research on vipers, and a book was planned based on those talks (Biology of the Vipers, 2002). I was invited to present my long-term research on cottonmouths, *Agkistrodon piscivorus.* Long-term research often produces different results than short-term studies, but it was difficult to get funding for multi-year projects. Another negative is that long-term research typically results in fewer papers, which negatively affects a faculty's academic success. This is even though the work may be much more extensive. That is why such studies often end up in books where longer works can be published.

This conference was organized by Gordon Schuett and Mats Hoggren and held at Odalgarden Conference Centre near Uppsala, Sweden. This conference center was more like a very fancy resort, complete with a private lake, paddleboats, hiking trails and a very attentive staff. Not what most snake field biologists are used to! Our rooms were not ready yet when our bus from the airport arrived, so the staff pampered us with breakfast, magnificent coffee, and gave us full use of the facilities and run of the grounds. Many of us went around the lake hunting for snakes and I found an adder, *Vipera berus.* I'm not sure how many herpetologists were there but the program said nearly 100 snake biologists from

A European adder *(Vipera berus).*

20 countries had responded. The snakes around the facility had no chance to avoid being caught and photographed many times.

During the conference over the next three days, we heard talks on current research on both Old World and New World vipers. In addition to the paper sessions, we had an evening tour of Hammarby, the farm home of Carolus Linnaeus, who was the father of biological nomenclature. It was now owned by the University of Uppsala, and the managers brought out some of the original type specimens (first ones ever collected) of various vipers like the timber rattlesnake, *Crotalus horridus,* and the southern copperhead, *Agkistrodon contortrix,* which were really cool to see. They also had an actor portray Linnaeus and come out and talk to the group. He berated us for not having published as many books as he had!

After the meeting, the organizers took us to an island in the Baltic Archipelago, in the Baltic Sea. My field notes don't say, but it likely was the island of Oja. It was cloudy and cold so, although lots of folks looked all around, we didn't find either smooth snakes *(Coronella austriaca)* or vipers as they were too smart to be out. The next day, one of the local herpetologists (I didn't write down who) took us to look for grass snakes, *Natrix natrix,* the most common of the three snakes found in Sweden. He told me that their numbers were going down because of environmental laws requiring compost bins to be more secure. The snakes used to lay their eggs in the compost but could no longer get into the boxes to do that. They used the piles of rotting vegetation along lake shores, which heated up and incubated the eggs. It was so cool there in the summer that the snakes needed an artificial heat source to get the eggs to hatch before autumn.

The weather turned on us again and we did not get any snakes, just frogs. I had caught *Natrix* back at the venue, so the only species of snake I hadn't seen was the smooth snake. Our guide suggested the best place to find them was on the island of Oland, so Rich Seigel and I rented a car and headed there. We took a long bridge over to this large island.

Really a different place! There were walls of large stones everywhere, some still functioning as boundary lines and cattle pins, but most had fallen down over hundreds of years. Windmills were abundant too. As we drove, we saw stone cairns, fortresses, house foundations and burial stones. We had been told to go to the largest stone circle, Ismantorp Fortress, which was hidden in the Mittlandsskogen Forest. We found a sign directing us to it, and after a short walk through the forest, we found the fortress in a large open area. It was a large circle of huge stones with a sloping base angling up to the top, which was 20 feet high. It had a couple entrances, but all the internal rooms were now just outlined with only a few rocks. As we walked along the outer wall, we caught several smooth snakes. Ron Gutberlet from Tyler and Travis LeDuc from Austin showed up as we were leaving. They too had been instructed that this was the best place for these snakes. Yay, I got to see all three species of European snakes!

As I indicated, scientific meetings vary in size and length of time; however, they ultimately have two functions: to allow the dissemination of the most recent research sooner than by publication and to get participants together to brainstorm. I always gained more knowledge during informal interactions with other scientists during coffee breaks, meals, and planned outings in the evenings or after the meetings. The scientific talks did get me excited about getting back and continuing my studies. Over the years, the number of scientific meetings has proliferated. Universities have also looked at hosting conferences as a money-making process, instead of supporting them. Thus, most biologists have to pick and choose which meetings to attend. I obviously like the ones that feature more field trips!

REFERENCE LIST

Preface

Louv, Richard (2008). *Last Child in the Woods.* New York, NY: Workman Publishing.

Hecht, Bessie (1956). *All About Snakes.* New York, NY: Random House.

Schmidt, Karl P. and Davis, D. Dwight (1941). *Field Book of Snakes of the United States and Canada.* New York, NY: G.P. Putnam's Sons.

Chapter 1

Taylor, Edward Harrison (1975). *Recollections of an herpetologist.* Lawrence, OK: Museum of Natural History, University of Kansas.

Pough, Harvey F. (1976). Multiple Cryptic Effects of Crossbanded and Ringed Patterns of Snakes. *Copeia,* 1976(4), 834-836.

Rossman, Douglas A., Ford, Neil B. and Seigel, Richard A. (1996). *The Garter Snakes: Evolution and Ecology*. Norman, OK: University of Oklahoma Press.

Gehlbach, Frederick (1981). *Mountain Islands and Desert Seas: A Natural History of the U.S.-Mexican Borderlands*. College Station, TX: Texas A&M University Press.

Darwin, Charles (1972). *The Voyage of the Beagle*. New York, NY: Bantam Books.

Wallace, Alfred R. (1892). *Travels on the Amazon and Rio Negro*. London: Ward, Lock, Bowden and Co.

Belt, Thomas (1874). *The Naturalist in Nicaragua*. London: John Murray.

Chapter 2

Lillywhite, H.B. (1982). Tracking as an aid in ecological studies of snakes. In N.J. Scott (ed.), Herpetological Communities: a Symposium of the Society for the Study of Amphibians and Reptiles and the Herpetologists' League, August 1977. U.S. Fish and Wildlife *Wildlife Research Report* 13, pp. 181-191.

Chapter 3

Dundee, Harold A. and Miller, M.C. III (1968). Aggregative behavior and habitat conditioning by the prairie ringneck snake, *Diadophis punctatus arnyi*. *Tulane Studies in Zoology and Botany* 15(2), 41-58.

Penfound, William T. (1966). *1966 Story of Oliver's Woods*. Biodiversity Information and Data. https://biosurvey.ou.edu/oliver/1966story.html

Chapter 4

Henderson, Robert W., Schwartz, Albert and Noeske-Hallin, Teresa A. (1987). Food Habits of Three Colubrid Tree Snakes (Genus *Uromacer*) on Hispaniola. *Herpetologica*, 43(2), 241-248.

Chapter 5

Tennant, Alan and Werler, John E. (1982). *Snakes of Texas: A Field Guide*. Austin, TX: University of Texas Press.

Chapter 9

Daltry, Jennifer (Oct. 8, 2012). *A race against the odds*. Fauna & Flora International. https://www.fauna-flora.org/news/a-race-against-the-odds

St. Eustatius National Parks Foundation (STENAPA). https://www.statiapark.org/

Chapter 10

Carpenter, C. (1952). Comparative Ecology of the Common Garter Snake (*Thamnophis s. sirtalis*), the Ribbon Snake (*Thamnophis s. sauritus*), and Butler's Garter Snake (*Thamnophis butleri*) in Mixed Populations. *Ecological Monographs, 22*(4), 236-258.

Rossman, Douglas A., Ford, Neil B. and Seigel, Richard A. (1996). *The Garter Snakes: Evolution and Ecology*. Norman, OK: University of Oklahoma Press.

Echelle, Alice Fitch and Stewart, Margaret M. (2000). Henry S. Fitch. *Copeia, 2000*(3), 891-900.

Chapter 11

Rivas, Jesús A. (2020). *Anaconda: The Secret Life of the World's Largest Snake*. New York, NY: Oxford University Press.

Chapter 12

Ford, N. B. and Ford, D. F. (2002). Notes on the ecology of the South American water snake *Helicops angulatus* (Squamata: Colubridae) in Nariva Swamp, Trinidad. *Caribbean Journal of Science,* 38: 129-132.

Chapter 13

O'Donnell, R. P., Ford, N. B., Shine, R., and Mason, R. T. (2004). Male red-sided garter snakes, *Thamnophis sirtalis parietalis,* determine female mating status from pheromone trails. *Animal Behaviour, 68*(4), 677-683.

Chapter 16

Sparkman, A. M., Byars, D., Ford, N. and Bronikowski, A. (2010). The role of insulin-like growth factor-1 (IGF-1) in growth and reproduction in female brown house snakes *(Lamprophis fuliginosus). General and Comparative Endocrinology,* 168:408-414.

Byars, D. J., Ford, N. B., Sparkman, A. M. and Bronikowski, A. M. (2010). Influences of Diet and Family on Age of Maturation in Brown House Snakes, *Lamprophis fuliginosus. Herpetologica*, 66:456-463.

Chapter 17

Schuett, Gordon W. (2002). *Biology of the Vipers.* Eagle Mountain, Utah: Eagle Mountain Publishing.

DR. NEIL FORD

Dr. Neil Ford, a professor emeritus at The University of Texas at Tyler, earned his Ph.D. in zoology in 1979 from Miami University in Oxford, Ohio, and has studied snakes for over 50 years. His primary interest has been in life history trade-offs, i.e. such things as factors that affect the number and size of offspring. He collected and brought back to his laboratory snakes from numerous countries that had particular traits, which allowed him to test his hypothesizes. He has written over 70 peer-reviewed scientific papers, 16 popular articles and 11 book chapters on that research.

Along with Dr. Doug Rossman and Dr. Rich Seigel, he coauthored the book, "The Garter Snakes, Evolution and Ecology," published by the University of Oklahoma Press. This book details the taxonomy, behavior and ecology of the 30 species that make up this extremely successful group of snakes. He organized six national and international symposia in herpetology, and helped found the Snake Ecology Group, which brought together experts from around the world to share their research.

In 2015, along with Dr. Srini Kamphampati and Dr. Lance Williams, he started the UT Tyler Center For Environment, Biodiversity and Conservation. He retired in 2020 after 40 years in the UT Tyler Department of Biology.

Made in the USA
Las Vegas, NV
03 November 2026

57422980R00133